FAMILIAR LECTURES ON SCIENTIFIC SUBJECTS

FAMILIAR LECTURES ON SCIENTIFIC SUBJECTS

BY SIR JOHN F. W. HERSCHEL, BART., K. H.;

M.A.; D.C.L.; F.R.S. L. AND E.; HON. M.R.I.A; F.R.A.S.; M.C.U.P.S.
MEMBER OF THE INSTITUTE OF FRANCE; AND
CORRESPONDENT, ASSOCIATE, HONORARY OR ORDINARY MEMBER OF VARIOUS OTHER ACADEMIES AND INSTITUTIONS

NEW YORK:
GEORGE ROUTLEDGE & SONS,
416 Broome Street.
PHILADELPHIA: J. B. LIPPINCOTT & Co.
LONDON:
ALEXANDER STRAHAN & Co., AND GEORGE ROUTLEDGE & SONS.
1871.

PREFACE.

THE three first lectures in the following collection—on Earthquakes and Volcanoes, on the Sun, and on Comets—were delivered by the Author to a village audience, in the school-house of the parish of Hawkhurst, in Kent, his place of residence. They were subsequently printed as contributions to the pages of that excellent and useful periodical, *Good Words*, in which they were followed by those "On the Weather and Weather Prophets," "On Celestial Measurings and Weighings," and "On Light," the latter assuming in its progress the dimensions of a little elementary treatise, adapted for the perusal of non-mathematical readers. On the completion of this, it was suggested by the publisher of that work to collect and reprint them together,—a proposal the more welcome, as it afforded an opportunity for bringing together several other pieces of a somewhat similar character, some of which, though not properly characterized as "Lectures," it seemed desirable to reproduce. More especially, it appeared to the Author an imperative duty to let no opportunity pass of recalling the attention of the public to the great question of the proposed abandonment of our national system of weights and measures, and adoption in its stead of the metrical system of the French, with its unit, the metre, in place of the English yard, which has been so actively, and, in his opinion, so mischievously urged on Parliament; the agitation in favour of which only sleeps for the present, in the view of allowing the public mind to familiarize itself with the idea under a Permissive Act,—to be assuredly brought forward again with renewed activity, and under a more intense and prolonged pressure (to be met by a more concentrated and determined resistance), on no distant occasion.

No apology will be considered necessary for reproducing the little piece "On the Absorption of Light," the thirteenth in order of this collection. Though it does not pretend to anticipate any of the later experimental researches, and the reasonings grounded on them for concluding the conversion of motion into heat, electricity, and magnetism,—it is, nevertheless, a step (though a small one) in that direction, by showing that a state of apparent rest in a material body is not incompatible with the internal propagation *ad infinitum* within it of movement impressed on it from without. It is very conceivable that the internal or atomic organization of ponderable matter may be such as to concentrate and localize, in its individual molecular groups, the broken-up and dispersed undulations caused by any external shock; and so preserve them from attaining that final state of complete mutual counteraction which is there contemplated.

Some slight alterations in the wording, and additions (not in every instance unimportant) to the matter of the several Essays here reproduced, have been made; as well as, here and there, some numerical corrections. In particular, the last little piece, "On the Estimation of Skill in Target-Shooting," has for one of its objects the correction of an error in one of the Author's former works, while, at the same time, calling attention to a subject generally, and even nationally, interesting.

COLLINGWOOD, *June* 8, 1866.

CONTENTS.

LECTURE I.

ABOUT VOLCANOS AND EARTHQUAKES.

PURPOSE in this Lecture to say something about volcanos and earthquakes. It is a subject I have thought a good deal about, and seen a little of, for though I have never been so fortunate as to have seen a volcano in eruption, or to have been shaken out of my bed by an earthquake, still I have climbed the cones of Vesuvius and Etna, hammer in hand and barometer on back, and have wandered over and geologized among, I believe, nearly all the principal scenes of extinct volcanic activity in Europe, those of Spain excepted.

(2.) Every one knows that a volcano is a mountain that vomits out fire, and smoke, and cinders, and melted lava, and sulphur, and steam, and gases, and all kinds of horrible things; nay, even sometimes mud, and boiling water, and fishes; and everybody has heard or read of the earth opening, and swallowing up man and beast, and houses and churches; and closing on them with a snap, and smashing them to pieces; and then perhaps opening again, and casting them out with a flood of dirty water from some river or lake that had been gulped down with them. Now, all this, and much more, is literally true, and has happened over and over again; and when we have imagined it all, we shall have formed a tolerably correct notion of some at least of these visitations. And perhaps some may have been tempted to ask why and how it is that God has permitted this fair earth to be visited with such destruction. It can hardly be for the sins of men: for when these things occur they involve alike the innocent and the guilty; and besides, the volcano and the earthquake were raging on this earth with as much, nay greater violence, thousands and thousands of years before man ever set his foot upon it. But perhaps, on the other hand, it may have occurred to some to ask themselves whether it is not just possible that these ugly affairs are sent among us for some beneficent purpose; or at all events that they may form part and parcel of some great scheme of providential arrangement which is at work for good, and not for ill. A ship sometimes strikes on a rock, and all on board perish; a railway train runs into another, or breaks down, and then wounds and contusions are the order of the day; but nobody doubts that navigation and railway communication are great blessings. None of the great natural provisions for producing good are exempt in their workings from producing occasional mischief. Storms disperse and dilute pestilential vapours, and lightnings decompose and destroy them; but both the one and the other often annihilate the works of man, and inflict upon him sudden death. Well, then, I think I shall be able to show that the volcano and the earthquake, dreadful as they are, as local and temporary visitations, are in fact unavoidable (I had almost said necessary) incidents in a vast system of action to which we owe the very ground we stand upon, the very land we inhabit, without which neither man, beast, nor bird would have a place for their existence, and the world would be the habitation of nothing but fishes.

(3.) Now, to make this clear, I must go a little out of my way and say something about the first principles of geology. Geology does not pretend to go back to the creation of the world, or concern itself about its primitive state, but it does concern itself with the changes it sees going on in it now, and with the evidence of a long series of such changes it can produce in the most unmistakable features of the structure of our rocks and soil, and the way in which they lie one on the other. *As to what we* SEE *going on.*—We see everywhere, and along every coast-line, the sea warring against the land, and everywhere overcoming it; wearing and eating it down, and battering it to pieces; grinding those pieces to powder; carrying that powder away, and spreading it out over its own bottom, by the continued effect of the tides and currents. Look at our chalk cliffs, which once, no doubt, extended across the Channel to the similar cliffs on the French coast. What do we see? Precipices cut down to the sea-beach, constantly hammered by the waves and constantly crumbling: the beach itself made of the flints outstanding after the softer chalk has been ground down and washed away; themselves grinding one another under the same ceaseless discipline; first rounded into pebbles, then worn into sand, and then carried out farther and farther down the slope, to be replaced by fresh ones from the same source.

(4.) Well: the same thing is going on *everywhere, round every coast* of Europe, Asia, Africa, and America. Foot by foot or inch by inch, month by month or century by century, *down everything* MUST *go.* Time is as nothing in geology. And what the sea is doing the rivers are helping it to do. Look at the sand-banks at the mouth of the Thames. What are they but the materials of our island carried out to sea by the stream? The Ganges carries away from the soil of India, and delivers into the sea, twice as much solid substance *weekly* as is contained in the great pyramid of Egypt. The Irawaddy sweeps off from Burmah 62 cubic feet of earth in every second of time on an average, and there are 86,400 seconds in every day, and 365 days in every year; and so on for the other rivers. What has become of all that great bed of chalk which once covered all the weald of Kent, and formed a continuous mass from Ramsgate and Dover to Beechy Head, running inland to Madamscourt Hill and Seven Oaks? All clean gone, and swept out into the bosom of the Atlantic, and there forming other chalk-beds. Now, geology assures us, on the most conclusive and undeniable evidence, that ALL our present land, all our continents and islands, have been formed in this way out of the ruins of former ones. The old ones which existed at the beginning of things have all perished, and what we now stand upon has most assuredly been, at one time or other, perhaps many times, the bottom of the sea.

(5.) Well, then, there is power enough at work, and it has been at work long enough, utterly to have cleared away and spread over the bed of the sea all our present existing continents and islands, had they been placed where they are at the creation of the world; and from this it follows, as clear as demonstration can make it, that without *some* process of renovation or restoration to act in antagonism to this destructive work of old Neptune, there would not now be remaining a foot of dry land for living thing to stand upon.

(6.) Now, what *is* this process of restoration? Let the volcano and the earthquake tell their tale. Let the earthquake tell how, within the memory of man—under the eyesight of eye-witnesses, one of whom (Mrs Graham) has described the fact—the whole coast line of Chili, for 100 miles about Valparaiso, with the mighty chain of the Andes—mountains to which the Alps shrink into insignificance—was hoisted at one blow (in a single night, Nov. 19, A.D. 1822) from two to seven feet above its former level, leaving the beach *below* the old low water-mark high and dry; leaving the shell-fish sticking on the rocks out of reach of water; leaving the seaweed rotting in the air, or rather drying up to dust under the burning sun of a coast where rain never falls. The ancients had a fable of Titan hurled from heaven and buried under Etna, and by his struggles causing the earthquakes that desolated Sicily. But here we have an exhibition of Titanic forces on a far mightier scale. One of the Andes upheaved on this occasion was the gigantic mass of Aconcagua, which overlooks Valparaiso. To bring home to the mind the conception of such an effort, we must form a clear idea of what sort of mountain this is. It is nearly 24,000 feet in height. Chimborazo, the loftiest of the volcanic cones of the Andes, is lower by 2500 feet; and yet Etna, with Vesuvius at the top of it, and another Vesuvius piled on that, *would little more than surpass the midway height of the snow-covered portion of* that *cone*, which is one of the many chimneys by which the hidden fires of the Andes find vent. On the occasion I am speaking of, at least 10,000 square miles of country were estimated as having been upheaved, and the upheaval was not confined to the land, but extended far away to sea, which was proved by the soundings off Valparaiso, and along the coast, having been found considerably shallower than they were before the shock.

(7.) Again, in the year 1819, in an earthquake in India, in the district of Cutch, bordering on the Indus, a tract of country more than fifty miles long and sixteen broad was suddenly raised ten feet above its former level. The raised portion still stands up above the unraised, like a long perpendicular wall, which is known by the name of the "Ullah Bund," or "God's Wall." And again, in 1538, in that convulsion which threw up the Monte Nuovo (New Mountain), a cone of ashes 450 feet high, in a single night; the whole coast of Pozzuoli, near Naples, was raised twenty feet above its former level, and remains so permanently upheaved to this day. And I could mention innumerable other instances of the same kind.*

(8.) This, then, is the manner in which the earthquake does its work; *and it is always at work.* Somewhere or other in the world, there is perhaps not a day, certainly not a month, without an earthquake. In those districts of South and Central America, where the great chain of volcanic cones is situated—Chimborazo, Cotopaxi, and a long list with names unmentionable, or at least unpronounceable—the inhabitants no more think of counting earthquake shocks than we do of counting showers of rain. Indeed, in some places along that coast, a shower is a greater rarity. Even in our own island, near Perth, a year seldom passes without a shock, happily, within the records of history, never powerful enough to do any mischief.

(9.) It is not everywhere that this process goes on by fits and starts. For instance, the northern gulfs, and borders of the Baltic Sea, are steadily shallowing: and the whole mass of Scandinavia, including Norway, Sweden, and Lapland, is rising out of the sea at the average rate of about two feet per century. But as this fact (which is perfectly well established by reference to ancient high and low water-marks) is not so evidently connected with the action of earthquakes, I shall not further refer to it just now. All that I want to show is,

* Not that earthquakes *always raise* the soil; there are plenty of instances of subsidence, etc.

that there is a great cycle of changes going on, in which the earthquake and volcano act a very conspicuous part, *and that part a restorative and conservative one;* in opposition to the steadily destructive and levelling action of the ocean waters.

(10.) How this can happen; what can be the origin of such an enormous power thus occasionally exerting itself, will no doubt seem very marvellous—little short, indeed, of miraculous intervention—but the mystery, after all, is not quite so great as at first it seems. We are permitted to look a little way into these great secrets of nature; not far enough, indeed, to clear up every difficulty, but quite enough to penetrate us with admiration of that wonderful system of counterbalances and compensations; that adjustment of causes and consequences, by which, throughout all nature, evils are made to work their own cure; life to spring out of death; and renovation to tread in the steps and efface the vestiges of decay.

(11.) The key to the whole affair is to be found in the central heat of the earth. This is no scientific dream, no theoretical notion, but a fact established by direct evidence up to a certain point, and standing out from plain facts as a matter of unavoidable conclusion, in a hundred ways.

(12.) We all know that when we go into a cellar out of a summer sun, it feels *cool;* but when we go into it out of a wintry frost it is *warm.* The fact is, that a cellar, or a well, or any pit of a moderate depth, has always, day and night, summer and winter, the same degree of warmth, the same *temperature*, as it is called: and *that* always and everywhere is the same, or nearly the same, as the average warmth of the climate of the place. Forty or fifty feet deep in the ground, a thermometer here, in this spot,* would always mark the same degree, 49° that is, or seventeen degrees above the freezing point. Under the equator, at the same depth, it always stands at 84°, which is our *hot summer* heat, but which there is the average heat of the whole year. And this is so everywhere. Just at the surface, or a few inches below it, the ground is warm in the daytime, cool at night: at two or three feet deep the difference of day and night is hardly perceptible, but that of summer and winter is considerable. But at forty or fifty feet this difference also disappears, and you find a perfectly fixed, uniform degree of warmth, day and night; summer and winter; year after year.

(13.) But when we go deeper, as, for instance, down into mines or coal-pits, this one broad and general fact is always observed,—everywhere, in all countries, in all latitudes, in all climates, wherever there are mines, or deep subterranean caves,—the deeper you go, the hotter the earth is found to be. In one and the same mine, each particular depth has its own particular degree of heat, which never varies: but the lower always the hotter; and that not by a trifling, but what may well be called an astonishingly rapid rate of increase,—about a degree of the thermometer additional warmth for every 90 feet of additional depth, which is about 58° per mile!—so that,

* At Hawkhurst in Kent.

if we had a shaft sunk a mile deep, we should find in the rock a heat of 105°, which is much hotter than the hottest summer day ever experienced in Englånd.

(14.) It is not everywhere, however, that it is worth while to sink a shaft to any great depth; but borings for water (in what are called Artesian wells) are often made to enormous depths, and the water always comes up hot; and the deeper the boring, the hotter the water. There is a very famous boring of this sort in Paris, at La Grenelle. The water rises from a depth of 1794 feet, and its temperature is 82° of our scale, which is almost that of the equator. And, again, at Salzwerth, in Oeynhausen, in Germany, in a boring for salt-springs 2144 feet deep, the salt water comes up with a still higher heat, viz., 91°. Then, again, we have natural hot-water springs, which rise, it is true, from depths we have no means of ascertaining; but which, from the earliest recorded times, have always maintained the same heat. At Bath, for instance, the hottest well is 117° Fahr. On the Arkansas River, in the United States, is a spring of 180°; which is scalding hot; and that out of the neighbourhood of any volcano.

(15.) Now, only consider what sort of a conclusion this lands us in. This globe of ours is 8000 miles in diameter; a mile deep on its surface is a mere scratch. If a man had twenty greatcoats on, and I found under the first a warmth of 60° above the external air, I should expect to find 60° more under the second, and 60° more under the third, and so on; and, within all, *no man*, but a mass of red-hot iron. Just so with the outside crust of the earth. Every mile thick is such a greatcoat, and at 20 miles depth, according to this rate, the ground must be fully red-hot; and at no such very great depth beyond, either the whole must be melted, or only the most infusible and intractable kinds of material, such as our fire-clays and flints, would present some degree of solidity.

(16.) In short, what the icefloes and icebergs are to the polar seas, so we shall come to regard our continents and mountain-ranges in relation to the ocean of melted matter beneath. I do not mean to say there is no solid central mass; there may be one, or there may not, and, upon the whole, I think it likely enough that there is—kept solid, in spite of the heat, by the enormous pressure; but that has nothing to do with my present argument. All that I contend for is this,—Grant me a sea of liquid fire, on which we are all floating,—land and sea; for the bottom of the sea, anyhow, will not come nearly down to the lava level. The sea is probably nowhere more than five or six miles deep, which is far enough above that level to keep its bed from becoming red-hot.

(17.) Well, now, the land is perpetually wearing down, and the materials being carried out to sea. The coat of heavier matter is thinning off towards the land, and thickening over all the bed of the sea. What must happen? If a ship float even on her keel, transfer weight from the starboard to the larboard side, will she continue to float even? No, certainly. She will heel over to larboard. Many a good ship has gone to the bottom in this way. If the continents be lightened, they will rise; if the bed of the sea receive additional weight, it will sink. The bottom of the Pacific *is* sinking, in point of fact. Not that the Pacific is becoming *deeper*. This seems a paradox; but it is easily explained. The whole bed of the sea is in the act of being pressed down *by the laying on of new solid substance over its bottom.* The new bottom then is laid upon the old, and so the actual bed of the ocean remains at or nearly at the same distance from the surface water. But what becomes of the islands? They form part and parcel of the old bottom; and Dr Darwin has shown, by the most curious and convincing proofs, that they *are sinking*, and *have been sinking for ages*, and are only kept above water—by what, think you? By the labours of the coral insects, which always build up to the surface!

(18.) It is impossible but that this increase of pressure in some places and relief in others must be very unequal in their bearings. So that at some place or other this solid floating crust must be brought into a state of strain, and if there be a weak or a soft part, a crack will at last take place. When this happens, down goes the land on the heavy side, and up on the light side. Now this is exactly what took place in the earthquake which raised the Ullah Bund in Cutch. I have told you of a great crack drawn across the country, not far from the coast line; the inland country rose ten feet, but much of the sea-coast, and probably a large tract in the bed of the Indian Ocean, sank considerably below its former level. And just as you see when a crack takes place in ice, the water oozes up; so this kind of thing is always, or almost always, followed by an upburst of the subterranean fiery matter. The earthquake of Cutch was terminated by the outbreak of a volcano at the town of Bhooi, which it destroyed.

(19.) Now where, following out this idea, should we naturally expect such cracks and outbreaks to happen? Why, of course, along those lines where the relief of pressure on the land side is the greatest, and also its increase on the sea side; that is to say, along or in the neighbourhood of the sea-coasts, where the destruction of the land is going on with most activity. Well, now, it is a remarkable fact in the history of volcanos, that there is hardly an instance of an *active* volcano at any considerable distance from the sea-coast. All the great volcanic chain of the Andes is close to the western coast line of America. Etna is close to the sea; so is Vesuvius; Teneriffe is very near the African coast; Mount Erebus is on the edge of the great Antarctic continent. Out of 225 volcanos which are known to have been in actual eruption over the whole earth within the last 150 years, I remember only a single instance of one more than 320 miles from the sea, and even *that* is on the edge of the Caspian, the largest of all the inland seas—I mean Mount Demawend in Persia.

(20.) Suppose from this, or from any other cause, a crack to take place in the solid crust of the earth. Don't imagine that the melted matter below will simply ooze up quietly, as water does from under an ice-crack. No such thing. There is an element in the case we have not considered: steam and condensed gases. We all know what happens when a crack takes place in a high-pressure steam-boiler, with what violence the contents escape, and what havoc takes place. Now there is no doubt that among the minerals of the subterranean world, there is water in abundance, and sulphur, and many other vaporizable substances, all kept subdued and repressed by the enormous pressure. Let this pressure be relieved, and forth they rush, and the nearer they approach the surface the more they expand, and the greater is the explosive force they acquire; till at length, after more or fewer preparatory shocks, each accompanied with progressive weakening of the overlying strata, the surface finally breaks up, and forth rushes the imprisoned power, with all the awful violence of a volcanic eruption.

(21.) Certainly a volcano does seem to be a very bad neighbour; and yet it affords a compensation in the extraordinary richness of the volcanic soil, and the fertilizing quality of the ashes thrown out. The flanks or Somma (the exterior crater of Vesuvius) are covered with vineyards producing wonderful wine, and whoever has visited Naples, will not fail to be astonished at the productiveness of the volcanized territory as contrasted with the barrenness of the limestone rocks bordering on it. There you will see the amazing sight (as an English farmer would call it) of a triple crop growing at once on the same soil; a vineyard, an orchard, and a cornfield all in one. A magnificent wheat crop, five or six feet high, overhung with clustering grape-vines swinging from one apple or pear tree to another in the most luxuriant festoons! When I visited Somma, to see the country where the celebrated wine, the Lacryma Christi, is grown, it was the festival of the Madonna del Arco. Her church was crowded to suffocation with a hot and dusty assemblage of the peasantry. The fine impalpable volcanic dust was everywhere; in your eyes, in your mouth, begriming every pore; and there I saw what I shall never forget. Jammed among the crowd, I felt something jostling my legs. Looking down, and the crowd making way, I beheld a line of worshippers crawling on their hands and knees from the door of the church to the altar, licking the dusty pavement all the way with their tongues, positively applied to the ground and no mistake. No trifling dose of Lacryma would be required to wash down what they must have swallowed on that journey, and I have no doubt it was administered pretty copiously after the penance was over.

(22.) Now I come to consider the manner in which an earthquake is propagated from place to place; how it travels, in short. It runs along the earth precisely in the same manner, and according to the same mechanical laws as a wave along the sea, or rather as the waves of sound run along the air, but quicker. The earthquake which destroyed Lisbon ran out from thence, as from a centre, in all directions, at a rate averaging about twenty miles per minute, as far as could be gathered from a comparison of the times of its occurrence at different places; but there is little doubt that it must have been retarded by having to traverse all sorts of ground, for a blow or shock of any description is conveyed through the substance on which it is delivered with the rapidity of *sound in* that substance. Perhaps it may be new to many who hear me to be told that sound is conveyed by water, by stone, by iron, and indeed by everything, and at a different rate for each. In air it travels at the rate of about 1140 feet per second, or about 13 miles in a minute. In water much faster, more than four times as fast (4700 feet). In iron ten times as fast (11,400 feet), or about 130 miles in a minute, so that a blow delivered endways at one end of an iron rod, 130 miles long, would only reach the other after the lapse of a minute, and a pull at one end of an iron wire of that length, would require a minute before it would be felt at the other. But the substance of the earth through which the shock is conveyed is not only far less elastic than iron, but it does not form a coherent, connected body; it is full of interruptions, cracks, loose materials, and all these tend to deaden and retard the shock: and putting together all the accounts of all the earthquakes that have been exactly observed, their rate of travel may be taken to vary from as low as 12 or 13 miles a minute to 70 or 80: but perhaps the low velocities arise from oblique waves.

(23.) The way, then, that we may conceive an earthquake to travel is this,—I shall take the case which is most common, when the motion of the ground to-and-fro is horizontal. *How far* each particular spot on the surface of the ground is actually pushed from its place there is no way of ascertaining, since all the surrounding objects receive the same impulse almost at the same instant of time, but there are many indications that it is often several yards. In the earthquake of Cutch, which I have mentioned, trees were seen to flog the ground with their branches, which proves that their stems must have been jerked suddenly away for some considerable distance and as suddenly pushed back; and the same conclusion follows from the sudden rise of the water of lakes on the side where the shock reaches them, and its fall on the opposite side; the bed of the lake has been jerked away for a certain distance from under the water and pulled back.

(24.) Now, suppose a row of sixty persons, standing a mile apart from each other, in a straight line, in the direction in which the shock travels; at a rate, we will suppose, of sixty miles per minute: and let the ground below the first get a sudden and violent shove, carrying it a yard in the direction of the next. Since this shock will not reach the next till after the lapse of one second of time, it is clear that the space between the two will be shortened by a yard, and the ground—that is to say, not the mere loose soil on the surface, but the whole mass of solid rock below, down to an unknown depth—compressed, or driven into a smaller space. It is this compression that carries the shock forwards. The elastic force of the rocky matter, like a coiled spring acts both ways; it drives back the first man to his old place, and shoves the second a yard nearer to the third; and so on. Instead of men place a row of tall buildings, or columns, and they will tumble down in succession, the base flying forwards, and leaving the tops behind to drop on the soil on the side *from* which the shock came. This is just what was seen to happen in Messina in the great Calabrian earthquake. As the shock ran along the ground, the houses of the Faro were seen to topple down in succession; beginning at one end and running on to the other, as if a succession of mines had been sprung. In the earthquake in Cutch, a sentinel standing at one end of a long straight line of wall, saw the wall bow forward and recover itself; not all at once, but with a swell like a wave running all along it with immense rapidity. In this case it is evident that the earthquake wave must have had its front oblique to the direction of the wall (just as an obliquely-held ruler runs along the edge of a page of paper while it advances, like a wave of the sea, perpendicularly to its own length).

(25.) In reference to extinct volcanos, I may just mention that any one who wishes to see some of the finest specimens in Europe may do so by making a couple of days' railway travel to Clermont, in the department of the Puy de Dôme in France. There he will find a magnificent series of volcanic cones, fields of ashes, streams of lavas, and basaltic terraces or platforms, proving the volcanic action to have been continued for countless ages before the present surface of the earth was formed; and all so clear that he who runs may read their lesson. There can there be seen a configuration of surface quite resembling what telescopes show in the most volcanic districts of the moon. Let not my hearers be startled: half the moon's face is covered with unmistakable craters of extinct volcanos.

(26.) Many of the lavas of Auvergne and the Puy de

Dôme are basaltic; that is, consisting of columns placed close together; and some of the cones are quite complete, and covered with loose ashes and cinders, just as Vesuvius is at this hour.

(27.) In the study of these vast and awful phenomena we are brought in contact with those immense and rude powers of nature which seem to convey to the imagination the impress of brute force and lawless violence; but it is not so. Such an idea is not more derogatory to the wisdom and benevolence that prevails throughout all the scheme of creation than it is in itself erroneous. In their wildest paroxysms the rage of the volcano and the earthquake is subject to great and immutable laws: they feel the bridle and obey it. The volcano bellows forth its pent-up overplus of energy, and sinks into long and tranquil repose. The earthquake rolls away, and industry, that balm which nature knows how to shed over every wound, effaces its traces, and festoons its ruins with flowers. There is mighty and rough work to be accomplished, and it cannot be done by gentle means. It seems, no doubt, terrible, awful, perhaps harsh, that twenty or thirty thousand lives should be swept away in a moment by a sudden and unforeseen calamity; but we must remember that sooner or later every one of those lives must be called for, and it is by no means the most sudden end that is the most afflictive. It is well too that we should contemplate occasionally, if it were only to teach us humility and submission, the immense energies which are everywhere at work in maintaining the system of nature we see going on so smoothly and tranquilly around us, and of which these furious outbreaks, after all, are but minute, and for the moment unbalanced surpluses in the great account. The energy requisite to overthrow a mountain is as a drop in the ocean compared with that which holds it in its place, and makes it a mountain. Chemistry tells us that the forces constantly in action to maintain a single grain of water in its habitual state; when only partially and sparingly let loose in the form of electricity, would manifest themselves as a powerful flash of lightning.* And we learn from optical science that in even the smallest element of every material body, nay, even in *what we call* empty space, there are forces in perpetual action to which even such energies sink into insignificance. Yet, amid all this, nature holds her even course: the flowers blossom; animals enjoy their brief span of existence; and man has leisure and opportunity to contemplate and adore, secure of the watchful care which provides for his well-being at every instant that he is permitted to remain on earth.

ON THE HISTORY OF EARTHQUAKES AND VOLCANOS.

(28.) The first great earthquake of which any very distinct knowledge has reached us is that which occurred in the year 63 after our Saviour, which produced great destruction in the neighbourhood of Vesuvius, and shattered the cities of Pompeii and Herculaneum upon the Bay of Naples, though it did not destroy them. This earthquake is chiefly remarkable as having been

* Faraday: "Experimental Researches in Electricity," § 853.

the forerunner and the warning (if that warning could have been understood) of the first eruption of Vesuvius on record, which followed sixteen years afterwards in the year 79. Before that time none of the ancients had any notion of its being a volcano, though Pompeii itself is paved with its lava. The crater was probably filled, or at least the bottom occupied, by a lake; and we read of it as the stronghold of the rebel chief Spartacus, who, when lured there by the Roman army, escaped with his followers by clambering up the steep sides by the help of the wild vines that festooned them. The ground since the first earthquake in 63 had often been shaken by slight shocks, when at length, in August 79, they became more numerous and violent, and, on the night preceding the eruption, so tremendous as to threaten everything with destruction. A morning of comparative repose succeeded, and the terrified inhabitants of those devoted towns no doubt breathed more freely, and hoped the worst was over; when, about one o'clock in the afternoon, the Elder Pliny, who was stationed in command of the Roman fleet at Misenum in full view of Vesuvius, beheld a huge black cloud ascending from the mountain, which, "rising slowly always higher," at last spread out aloft like the head of one of those picturesque flat-topped pines which form such an ornament of the Italian landscape. The meaning of such a phenomenon was to Pliny and to every one a mystery. We know now too well what it imports, and they were not long left in doubt. From that cloud descended stones, ashes, and pumice; and the cloud itself lowered down upon the surrounding country, involving land and sea in profound darkness, pierced by flashes of fire more vivid than lightning. These, with the volumes of ashes that began to encumber the soil, and which covered the sea with floating pumice-stone; the constant heaving of the ground; and the sudden recoil of the sea, form a picture which is wonderfully well described by the Younger Pliny. His uncle, animated by an eager desire to know what was going on, and to afford aid to the inhabitants of the towns, made sail for the nearest point of the coast and landed; but was instantly enveloped in the dense sulphureous vapour that swept down from the mountain, and perished miserably.

(29.) It does not seem that any *lava* flowed on that occasion. Pompeii was buried under the ashes; Herculaneum by a torrent of mud, probably the contents of the crater, ejected at the first explosion. This was most fortunate. We owe to it the preservation of some of the most wonderful remains of antiquity. For it is not yet much more than a century ago that, in digging a well at Portici near Naples, the Theatre of Herculaneum was discovered, some sixty feet under ground,—then houses, baths, statues, and, most interesting of all, a library, full of books; and those books still legible, and among them the writings of some ancient authors which had never before been met with, but which have now been read, copied, and published, while hundreds and hundreds, I am sorry to say, still remain unopened. Pompeii was not buried so deep; the walls of some of the buildings appeared among the modern vineyards; and led to excavations, which were easy, the ashes being light and loose. And there you now may walk through the streets, enter the houses, and find the skeletons of their inmates, some in the very act of trying to escape. Nothing can be more strange and striking.

(30.) Since that time Vesuvius has been frequently but very irregularly in eruption. The next after Pompeii was in the year 202, under Severus: and in 472 occurred an eruption so tremendous that all Europe was covered by the ashes, and even Constantinople thrown into alarm. This may seem to savour of the marvellous; but before I have done, I hope to show that it is not beyond what we know of the power of existing volcanos.

(31.) I shall not, of course, occupy attention with a history of Vesuvius, but pass at once to the eruption of 1779,—one of the most interesting on record, from the excellent account given of it by Sir William Hamilton, who was then resident at Naples as our Minister, and watched it throughout with the eye of an artist as well as the scrutiny of a philosopher.

(32.) In 1767, there had been a considerable eruption, during which Pliny's account of the great pine-like, flat-topped, spreading mass of smoke had been superbly exemplified; extending over the Island of Capri, which is twenty-eight miles from Vesuvius. The showers of ashes, the lava currents, the lightnings, thunderings, and earthquakes were very dreadful; but they were at once brought to a close when the mob insisted that the head of St Januarius should be brought out and shown to the mountain; and when this was done, all the uproar ceased on the instant, and Vesuvius became as quiet as a lamb!

(33.) He did not continue so, however, and it would have been well for Naples if the good Saint's head could have been permanently fixed in some conspicuous place in sight of the hill—for from that time till the year 1779 it never was quiet. In the spring of that year it began to pour out lava; and on one occasion, when Sir William Hamilton approached too near, the running stream was on the point of surrounding him; and the sulphureous vapour cut off his retreat, so that his only mode of escape was to walk across the lava, which, to his astonishment, and, no doubt, to his great joy, he found accompanied with no difficulty, and with no more inconvenience than what proceeded from the radiation of heat on his legs and feet from the scoriæ and cinders with which the external crust of the lava was loaded; and which in great measure intercepted and confined the glowing heat of the ignited mass below.

(34.) In such cases, and when cooled down to a certain point, the motion of the lava-stream is slow and creeping; rather rolling over itself than flowing like a river; the top becoming the bottom, owing to the toughness of the half-congealed crust. When it issues, however, from any accessible vent, it is described as perfectly liquid, of an intense white heat, and spouting or welling forth with extreme rapidity. So Sir Humphry Davy described it in an eruption at which he was present;

and so Sir William Hamilton, in the eruption we are now concerned with, saw it "bubbling up violently" from one of its fountains on the slope of the volcano, "with a hissing and crackling noise, like that of an artificial firework; and forming, by the continual splashing up of the vitrified matter, a sort of dome or arch over the crevice from which it issued," which was all, internally, "red-hot like a heated oven."

(35.) However, as time went on, this quiet mode of getting rid of its contents would no longer suffice, and the usual symptoms of more violent action—rumbling noises and explosions within the mountain; puffs or smoke from its crater, and jets of red-hot stones and ashes—continued till the end of July, when they increased to such a degree as to exhibit at night the most beautiful firework imaginable. The eruption came to its climax from the 5th to the 10th of August, on the former of which days, after the ejection of an enormous volume of white clouds, piled like bales of the whitest cotton, in a mass exceeding four times the height and size of the mountain itself; the lava began to overflow the rim of the crater, and stream in torrents down the steep slope of the cone. This was continued till the 8th, when the great mass of the lava would seem to have been evacuated, and no longer repressing by its weight the free discharge of the imprisoned gases, allowed what remained to be ejected in fountains of fire, carried up to an immense height in the air. The description of one of these I must give in the picturesque and vivid words of Sir William Hamilton himself. "About nine o'clock," he says, on Sunday the 8th of August, " there was a loud report, which shook the houses at Portici and its neighbourhood to such a degree, as to alarm the inhabitants and drive them out into the streets. Many windows were broken, and as I have since seen, walls cracked by the concussion of the air from that explosion. . . . . In one instant a fountain of liquid transparent fire began to rise, and gradually increasing, arrived at so amazing a height, as to strike every one who beheld it with the most awful astonishment. I shall scarcely be credited when I assure you that, to the best of my judgment, the height of this stupendous column of fire could not be less than three times that of Vesuvius itself; which, you know, rises perpendicularly near 3700 feet above the level of the sea." (The height by my own measurement in 1824 is 3920 feet). "Puffs of smoke, as black as can possibly be imagined, succeeded one another hastily, and accompanied the red-hot, transparent, and liquid lava, interrupting its splendid brightness here and there by patches of the darkest hue. Within these puffs of smoke, at the very moment of their emission from the crater, I could perceive a bright but pale electrical fire playing about in zigzag lines. The liquid lava, mixed with scoriæ and stones, after having mounted, I verily believe at least 10,000 feet, falling perpendicularly on Vesuvius, covered its whole cone, part of that of Somma, and the valley between them. The falling matter being nearly as vivid and inflamed as that which was continually issuing fresh from the crater, formed with it one complete body of fire, which could not be less than two miles and a half in breadth, and of the extraordinary height above mentioned; casting a heat to the distance of at least six miles around it. The brushwood of the mountain of Somma was soon in a flame, which, being of a different tint from the deep red of the matter thrown out from the volcano, and from the silvery blue of the electrical fire, still added to the contrast of this most extraordinary scene. After the column of fire had continued in full force for nearly half an hour, the eruption ceased at once, and Vesuvius remained sullen and silent.'

(36.) The lightnings here described arose evidently in part from the chemical activity of gaseous decompositions going forward, in part to the friction of steam, and in part from the still more intense friction of the dust, stones, and ashes encountering one another in the air, in analogy to the electric manifestations which accompany the dust storms in India.

(37.) To give an idea of the state of the inhabitants of the country when an explosion is going on, I will make one other extract:—"The mountain of Somma, at the foot of which Ottaiano is situated, hides Vesuvius from its sight: so that till the eruption became considerable it was not visible to them. On Sunday night, when the noise increased, and the fire began to appear above the mountain of Somma, many of the inhabitants of the town flew to the churches; and others were preparing to quit the town, wnen a sudden violent report was heard, soon after which they found themselves involved in a thick cloud of smoke and minute ashes; a horrid clashing noise was heard in the air; and presently fell a deluge of stones and large scoriæ, some of which scoriæ were of the diameter of seven or eight feet, and must have weighed more than one hundred pounds before they were broken by their falls, as some of the fragments of them which I picked up in the streets still weighed upwards of sixty pounds. When these large vitrified masses either struck against each other in the air or fell on the ground they broke in many pieces, and covered a large space around them with vivid sparks of fire, which communicated their heat to everything that was combustible. In an instant the town and country about it was on fire in many parts; for in the vineyards there were several straw-huts, which had been erected for the watchmen of the grapes, all of which were burnt. A great magazine of wood in the heart of the town was all in a blaze: and had there been much wind, the flames must have spread universally, and all the inhabitants would have infallibly been burnt in their houses, for it was impossible for them to stir out. Some who attempted it with pillows, tables, chairs, tops of wine casks, etc., on their heads, were either knocked down or driven back to their close quarters, under arches and in the cellars of the houses. Many were wounded, but only two persons have died of the wounds they received from this dreadful volcanic shower. To add to the horror of the scene, incessant volcanic lightning was writhing about the black cloud that surrounded them, and the sulphureous smell and heat would scarcely allow them to draw their breath.*

(38.) The next volcano I shall introduce is Ætna, the grandest of all our European volcanos. I ascended it in 1824, and found its height by a very careful barometric measurement to be 10,772 feet above the sea, which, by the way, agrees within some eight or ten feet with Admiral Smyth's measurement.

(39.) The scenery of Ætna is on the grandest scale. Ascending from Catania you skirt the stream of lava which destroyed a large part of that city in 1669, and which ran into the sea, forming a jetty or breakwater that now gives Catania what it never had before, the advantage of a harbour. There it lies as hard, rugged, barren, and fresh-looking as if it had flowed but yesterday. In many places it is full of huge caverns; great air-bubbles, into which one may ride on horseback (at least large enough) and which communicate, in a succession of horrible vaults, where one might wander and lose one's self without hope of escape. Higher up, near Nicolosi, is the spot from which that lava flowed. It is marked by two volcanic cones, each of them a considerable mountain, called the Monti Rossi, rising 300 feet above the slope of the hill, and which were thrown up on that occasion. Indeed, one of the most remarkable features of Ætna is that of its flanks bristling over with innumerable smaller volcanos. For the height is so great that the lava now scarcely ever rises to the top of the crater; for before that, its immense weight breaks through at the sides. In one of the eruptions that happened in the early part of this century, I forget the date, but I think it was in 1819, and which was described to me on the spot by an eye-witness—the Old Man of the Mountain, Mario Gemellaro—the side of Ætna was rent by a great fissure or crack, beginning near the top, and throwing out jets of lava from openings fourteen or fifteen in number all the way down, so as to form a row of fiery fountains rising from different levels, and all ascending nearly to the same height: thereby proving them all to have originated in the great internal cistern as it were, the crater being filled up to the top level.

(40.) From the summit of Ætna extends a view of extraordinary magnificence. The whole of Sicily lies at your feet, and far beyond it are seen a string of lesser volcanos; the Lipari Islands, between Sicily and the Italian coast; one of which, Stromboli, is always in eruption, unceasingly throwing up ashes, smoke, and liquid fire.

(41.) But I must not linger on the summit of Ætna. We will now take a flight thence, all across Europe, to Iceland—a wonderful land of frost and fire. It is full of volcanos, one of which, Hecla, has been twenty-two times in eruption within the last 800 years. Besides Hecla, there are five others, from which in the same period twenty eruptions have burst forth, making about one every twenty years. The most formidable of these was that which happened in 1783, a year also memorable as that of the terrible earthquake in Calabria. In May of that year, a bluish fog was observed over the mountain called Skaptar Jokul, and the neighbourhood was shaken by earthquakes. After a while a great pillar of smoke was observed to ascend from it, which darkened the whole surrounding district, and which descended in a whirlwind of ashes. On the 10th of May, innumerable fountains of fire were seen shooting up through the ice and snow which covered the mountain; and the principal river, called the Skapta, after rolling down a flood of foul and poisonous water, disappeared. Two days after, a torrent of lava poured down into the bed which the river had deserted. The river had run in a ravine, 600 feet deep and 200 broad. This the lava entirely filled; and not only so, but it overflowed the surrounding country, and ran into a great lake, from which it instantly expelled the water in an explosion of steam. When the lake was fairly filled, the lava again overflowed and divided into two streams, one of which covered some ancient lava fields; the other re-entered the bed of the Skapta lower down; and presented the astounding sight of a cataract of liquid fire pouring over what was formerly the waterfall of Stapafoss. This was the greatest eruption on record in Europe. It lasted in its violence till the end of August, and closed with a violent earthquake; but for nearly the whole year a canopy of cinder-laden cloud hung over the island; the Faroe Islands, nay, even Shetland and the Orkneys, were deluged with the ashes; and volcanic dust and a preternatural smoke, which obscured the sun, covered all Europe as far as the Alps, over which it could not rise. It has been surmised that the great Fireball of August 18, 1783, which traversed all England and the Continent, from the North Sea to Rome, by far the greatest ever known (for it was more than half a mile in diameter), was somehow connected with the electric excitement of the upper atmosphere produced by this enormous discharge of smoke and ashes. The destruction of life in Iceland was frightful: 9000 men, 11,000 cattle, 28,000 horses, and 190,000 sheep perished; mostly by suffocation. The lava ejected has been computed to have amounted in volume to more than twenty cubic miles.

(42.) We shall now proceed to still more remote regions, and describe, in as few words as may be, two immense eruptions,—one in Mexico, in the year 1759; the other in the island of Sumbawa in the Eastern Archipelago, in 1815.

(43.) I ought to mention, by way of preliminary, that almost the whole line of coast of South and Central America, from Mexico southwards as far as Valparaiso—that is to say, nearly the whole chain of the Andes—is one mass of volcanos. In Mexico and Central America there are two and twenty, and in Quito, Peru, and Chili, six and twenty more, in activity; and nearly as many more extinct ones, any one of which may at any moment break out afresh. This does not prevent the country from being inhabited, fertile, and well cultivated. Well: in a district of Mexico celebrated for the growth of the finest cotton, between two streams called Cuitimba and San Pedro, which furnished water for irrigation, lay the farm and homestead of Don Pedro de Jurullo, one of the richest and most fertile properties in that country. He was a thriving man, and lived in comfort as a large proprietor, little expecting the mischief that was to befall him. In June 1759, however, a subterranean noise was heard in this peaceful region. Hollow sounds of the most alarming nature were succeeded by frequent earthquakes, succeeding one another for fifty or sixty days; but they died away, and in the beginning of September everything seemed to have returned to its usual state of tranquillity. Suddenly, on the night of the 28th of September, the horrible noises recommenced. All the inhabitants fled in terror; and the whole tract of ground, from three to four square miles in extent, rose up in the form of a bladder to a height of upwards of 500 feet! Flames broke forth over a surface of more than half a square league, and through a thick cloud of ashes illuminated by this ghastly light, the refugees, who had ascended a mountain at some distance, could see the ground as if softened by the heat, and swelling and sinking like an agitated sea. Vast rents opened in the earth, into which the two rivers I mentioned precipitated themselves, but so far from quenching the fires, only seemed to make them more furious. Finally, the whole plain became covered with an immense torrent of boiling mud, out of which sprang thousands of little volcanic cones called Hornitos, or ovens. But the most astonishing part of the whole was the opening of a chasm vomiting out fire, and red-hot stones, and ashes, which accumulated so as to form "a range of six large mountain masses, one of which is upwards of 1600 feet in height above the old level, and which is now known as the volcano of Jorullo. It is continually burning; and for a whole year continued to throw up an immense quantity of ashes, lava, and fragments of rock. The roofs of houses at the town or village of Queretaro, upwards of 140 miles distant, were covered with the ashes. The two rivers have again appeared, issuing at some distance from among the hornitos, but no longer as sources of wealth and fertility, for they are scalding hot, or at least were so when Baron Humboldt visited them several years after the event. The ground even then retained a violent heat, and the hornitos were pouring forth columns of steam twenty or thirty feet high, with a rumbling noise like that of a steam-boiler.

(44.) The Island of Sumbawa is one of that curious line of islands which links on Australia to the south-eastern corner of Asia. It forms, with one or two smaller volcanic islands, a prolongation of Java, at that time, in 1815, a British possession, and under the government of Sir Stamford Raffles, to whom we owe the account of the eruption, and who took a great deal of pains to ascertain all the particulars. Java itself, I should observe, is one rookery of volcanos, and so are all the adjoining islands in that long crescent-shaped line I refer to.

(45.) On the Island of Sumbawa is the volcano of Tomboro, which broke out into eruption on the 5th of April in that year; and I can hardly do better than quote the account of it in Sir Stamford Raffles' own words:—

(46.) "Almost every one," says this writer, "is acquainted with the intermitting convulsions of Etna and Vesuvius as they appear in the descriptions of the poet, and the authentic accounts of the naturalist; but the most extraordinary of them can bear no comparison, in point of duration and force, with that of Mount Tomboro in the Island of Sumbawa! This eruption extended perceptible evidences of its existence over the whole of the Molucca Islands, over Java, a considerable portion of the Celebes, Sumatra, and Borneo, to a circumference of 1000 statute miles from its centre" (*i.e.*, to 1000 miles' *distance*), "by tremulous motions and the report of explosions. In a short time the whole mountain near the Sang'ir appeared like a body of liquid fire, extending itself in every direction. The fire and columns of flame continued to rage with unabated fury, until the darkness, caused by the quantity of falling matter, obscured it at about eight P.M. Stones at this time fell very thick at Sang'ir, some of them as large as two fists, but generally not larger than walnuts. Between nine and ten P.M., ashes began to fall, and soon after a violent whirlwind ensued, which blew down nearly every house of Sang'ir, carrying the roofs and light parts away with it. In the port of Sang'ir, adjoining Sumbawa, its effects were much more violent, tearing up by the roots the largest trees, and carrying them into the air, together with men, horses, cattle, and whatsoever came within its influence. This will account for the immense number of floating trees seen at sea. The sea rose nearly twelve feet higher than it had ever been known to do before, and completely spoiled the only small spots of rice land in Sang'ir, sweeping away houses and everything within its reach. The whirlwind lasted about an hour. No explosions were heard till the whirlwind had ceased at about eleven P.M. From midnight till the evening of the 11th, they continued without intermission; after that time their violence moderated, and they were heard only at intervals; but the explosions did not cease entirely until the 15th of July. Of all the villages round Tomboro, Tempo, containing about forty inhabitants, is the only one remaining. In Pekaté no vestige of a house is left; twenty-six of the people, who were at Sumbawa at the time, are the whole of the population who have escaped. From the best inquiries, there were certainly not fewer than 12,000 individuals in Tomboro and Pekaté at the time of the eruption, of whom five or six survive. The trees and herbage of every description, along the whole of the north and west of the peninsula, have been completely destroyed, with the exception of a high point of land near the spot where the village of Tomboro stood. At Sang'ir, it is added, the famine occasioned by this event was so extreme, that one of the rajah's own daughters died of starvation."

(47.) I have seen it computed that the quantity of ashes and lava vomited forth in this awful eruption would have formed three mountains the size of Mont Blanc, the highest of the Alps; and if spread over the surface of Germany, would have covered the whole of it two feet deep! The ashes did actually cover the whole island of Tombock, more than 100 miles distant, to that depth, and 44,000 persons there perished by starvation, from the total destruction of all vegetation.

(48.) The mountain Kirauiah in the island of Owyhee,

one of the Sandwich Isles, exhibits the remarkable phænomenon of a lake of molten and very liquid lava *always* filling the bottom of the crater, and always in a state of terrific ebullition : rolling to and fro its fiery surge and flaming billows—yet with this it is content, for it would seem that at least for a long time past there has been no violent outbreak so as to make what is generally understood by a volcanic eruption. Volcanic eruptions are almost always preceded by earthquakes, by which the beds of rock, that overlie and keep down the struggling powers beneath, are dislocated and cracked, till at last they give way, and the strain is immediately relieved. It is chiefly when this does not happen, when the force below is sufficient to heave up and shake the earth, but not to burst open the crust, and give vent to the lava and gases, that the most destructive effects are produced. The great earthquake of November 1, 1755, which destroyed Lisbon, was an instance of this kind, and was one of the greatest, if not the very greatest on record ; for the concussion extended over all Spain and Portugal—indeed, over all Europe, and even into Scotland—over North Africa, where in one town in Morocco 8000 or 10,000 people perished. Nay, its effects extended even across the Atlantic to Madeira, where it was very violent ; and to the West Indies. The most striking feature about this earthquake was its extreme suddenness. All was going on quite as usual in Lisbon the morning of that memorable day ; the weather fine and clear ; and nothing whatever to give the population of that great capital the least suspicion of mischief. All at once, at twenty minutes before ten A.M., a noise was heard like the rumbling of carriages under ground; it increased rapidly and became a succession of deafening explosions like the loudest cannon. Then a shock, which, as described by one writing from the spot, seemed to last but the tenth part of a minute; and down came tumbling palaces, churches, theatres, and every large public edifice, and about a third or a fourth part of the dwelling houses. More shocks followed in rapid succession, and in six minutes from the commencement 60,000 persons were crushed in the ruins! Here are the simple but expressive words of one J. Latham, who writes to his uncle in London. "I was on the river with one of my customers going to a village three miles off. Presently the boat made a noise as if on the shore or landing, though then in the middle of the water. I asked my companion if he knew what was the matter. He stared at me, and looking at Lisbon, we saw the houses falling, which made him say, 'God bless us, it is an earthquake!' About four or five minutes after, the boat made a noise as before; and we saw the houses tumble down on both sides of the river." They then landed and made for a hill; whence they beheld the sea (which had at first receded and laid a great tract dry) come rolling in, in a vast mountain wave fifty or sixty feet high, on the land, and sweeping all before it. Three thousand people had taken refuge on a new stone quay or jetty just completed at great expense. In an instant it was turned topsy-turvy; and the whole quay, and every person on it, with all the vessels moored to it, disappeared, and not a vestige of them ever appeared again. Where that quay stood, was afterwards found a depth of 100 fathoms (600 feet) water. It happened to be a religious festival, and most of the population were assembled in the churches, which fell and crushed them. That no horror might be wanting, fires broke out in innumerable houses where the wood-work had fallen on the fires; and much that the earthquake had spared was destroyed by fire. And then too broke forth that worst of all scourges, a lawless ruffian-like mob, who plundered, burned, and murdered in the midst of all that desolation and horror. The huge wave I have spoken of swept the whole coast of Spain and Portugal. Its swell and fall was ten or twelve feet at Madeira. It swept quite across the Atlantic, and broke on the shores of the West Indies. Every lake and firth in England and Scotland was dashed for a moment out of its bed, the water not partaking of the sudden *shove* given to the land, just as when you splash a flat saucerful of water, the water dashes over on the side *from* which the shock is given.

(49.) One of the most curious incidents in this earthquake was its effect on ships far out at sea, which would lead us to suppose that the immediate impulse was in the nature of a violent blow or thrust upwards, under the bed of the ocean. Thus it is recorded that this upward shock was so sudden and violent on a ship, at that time forty leagues from Cape St Vincent, that the sailors on deck were tossed up into the air to a height of eighteen inches. So also, on another occasion in 1796, a British ship eleven miles from land near the Philippine Islands was struck upwards from below with such force as to unship and split up the main-mast.

(50.) Evidences of a similar sudden and upward explosive action are of frequent occurrence among the extinct volcanos of Auvergne and the Vivarais, where in many instances the perforation of the granitic beds which form the basis or substratum of the whole country appears to have been effected at a single blow, accompanied with little evidence of disturbance of the surrounding rocks—much in the same way as a bullet will pass through a pane of glass without starring or shattering it. In such cases it would seem as if water in a liquid state had suddenly been let in through a fissure upon a most intensely heated and molten mass beneath, producing a violent but local explosion, so instantaneous as to break its way through the overlying rocks, without allowing time for them to bend or crumple, and so displace the surrounding masses.

(51.) The same kind of upward bounding movement took place at Riobamba in Quito in the great earthquake of February 4, 1797, which was connected with an eruption of the volcano of Tunguragua. That earthquake extended in its greatest intensity over an oval space of 120 miles from south to north, and 60 from east to west, within which space every town and village was levelled with the ground; but the total extent of surface shaken was upwards of 500 miles in one direction (from Puna to Popayan), and 400 in the other. Quero, Riobamba, and several other towns, were buried under fallen mountains, and in a very few minutes

30,000 persons were destroyed. At Riobamba, however, after the earthquake, a great number of corpses were found to have been tossed across a river, and scattered over the slope of a hill on the other side.

(52.) The frequency of these South American earthquakes is not more extraordinary than the duration of the shocks. Humboldt relates that on one occasion, when travelling on mule-back with his companion Bonpland, they were obliged to dismount in a dense forest, and throw themselves on the ground: the earth being shaken uninterruptedly for upwards of a quarter of an hour with such violence that they could not keep their legs.

(53.) One of the most circumstantially described earthquakes on record is that which happened in Calabria on the 5th of February 1783; I should say began then, for it may be said to have lasted four years. In the year 1783, for instance, 949 shocks took place, of which 501 were great ones, and in 1784, 151 shocks were felt, 98 of which were violent. The centre of action seemed to be under the towns of Monteleone and Oppido. In a circle twenty-two miles in radius round Oppido every town and village was destroyed within two minutes by the first shock, and within one of seventy miles' radius all were seriously shaken and much damage done. The whole of Calabria was affected, and even across the sea Messina was shaken, and a great part of Sicily.

(54.) There is no end of the capricious and out-of-the-way accidents and movements recorded in this Calabrian earthquake. The ground undulated like a ship at sea. People became actually sea-sick, and to give an idea of the undulation (just as it happens at sea), the scud of the clouds before the wind seemed to be fitfully arrested during the pitching movement when it took place in the same direction, and to redouble its speed in the reverse movement. At Oppido many houses were swallowed up bodily. Loose objects were tossed up several yards into the air. The flagstones in some places were found after a severe shock all turned bottom upwards. Great fissures opened in the earth, and at Terra Nova a mass of rock 200 feet high and 400 in diameter travelled four miles down a ravine. All landmarks were removed, and the land itself, in some instances, with trees and hedges growing on it, carried bodily away and set down in another place. Altogether about 40,000 people perished by the earthquakes, and some 20,000 more of the epidemic diseases produced by want and the effluvia of the dead bodies.

(55.) Volcanos occasionally break forth at the bottom of the sea, and, when this is the case, the result is usually the production of a new island. This, in many cases, disappears soon after its formation, being composed of loose and incoherent materials, which easily yield to the destructive power of the waves. Such was the case with the Island of Sabrina, thrown up, in 1811, off St Michaels, in the Azores, which disappeared almost as soon as formed, and in that of Pantellaria, on the Sicilian coast, which resisted longer, but was gradually washed into a shoal, and at length has, we believe, completely disappeared.* In numerous other instances, the cones of cinders and scoriæ, once raised, have become compacted and bound together by the effusion of lava, hardening into solid stone, and thus, becoming habitual volcanic vents, they continue to increase in height and diameter, and assume the importance of permanent volcanic islands. Such has been, doubtless, the history of those numerous insular volcanos which dot the ocean in so many parts of the world, such as Teneriffe, the Azores, Ascension, St Helena, Tristan d'Acunha, etc. In some cases the process has been witnessed from its commencement, as in that of two islands which arose in the Aleutian group, connecting Kamtschatka with North America, the one in 1796, the other in 1814, and which both attained the elevation of 3000 feet.

(56.) Besides these evident instances of eruptive action, there is every reason to believe that enormous floods of lava have been, at various remote periods in the earth's history, poured forth at the bottom of seas so deep as to repress, by the mere weight of water, all outbreak of steam, gas, or ashes; and reposing perhaps for ages in a liquid state, protected from the cooling action of the water on their upper surface by a thick crust of congealed stony matter, to have assumed a perfect level; and, at length, by slow cooling, taken on that peculiar columnar structure which we see produced in miniature in

* Such an event is at this moment in progress (March 1866), close to the island of Santorini, in the bay of Thera, in the Greek Archipelago: itself, with the adjacent Kaimeni Islands, products of the same kind.

starch by the contraction or shrinkage, and consequent splitting, of the material in drying; and resulting in those picturesque and singular landscape-features called basaltic colonnades: when brought up to day by sudden or gradual upheaval, and broken into cliffs and terraces by the action of waves, torrents, or weather. Those grand specimens of such colonnades which Britain possesses in the Giant's Causeway of Antrim, and the cave of Fingal in Staffa, for instance, are no doubt extreme outstanding portions of such a vast submarine lava-flood which at some inconceivably remote epoch occupied the whole intermediate space; affording the same kind of evidence of a former connexion of the coasts of Scotland and Ireland as do the opposing chalk cliffs of Dover and Boulogne of the ancient connexion of France with Britain. Here and there a small basaltic island, such as that of Rathlin, remains to attest this former continuity, and to recall to the contemplative mind that sublime antagonism between sudden violence and persevering effort, which the study of geology impresses in every form of repetition.

(57.) There exists a very general impression that earthquakes are preceded and ushered in by some kind of preternatural, and, as it were, expectant calm in the elements; as if to make the confusion and desolation they create the more impressive. The records of such visitations which we possess, however striking some particular cases of this kind may appear, by no means bear out this as a general fact, or go to indicate any particular phase of weather as preferentially accompanying their occurrence. This does not prevent, however, certain conjunctures of atmospheric or other circumstances from exercising a determining influence on the times of their occurrence. According to the view we have taken of their origin (viz., the displacement of pressure, resulting in a state of strain in the strata at certain points, gradually increasing to the maximum they can bear without disruption), it is the last ounce which breaks the camel's back. Great barometrical fluctuation, accumulating atmospheric pressure for a time over the sea, and relieving it over the land; an unusually high tide, aided by long-continued and powerful winds, heaping up the water; nay, even the tidal action of the sun and moon on the *solid* portion of the earth's crust,—all these causes, for the moment combining, may very well suffice to determine the instant of fracture, when the balance between the opposing forces is on the eve of subversion. The last-mentioned cause may need a few words of explanation. The action of the sun and moon, though it cannot produce a tide in the solid crust of the earth, *tends* to do so, and, were it fluid, *would* produce it. It therefore, in point of fact, does bring the solid portions of the earth's surface into a state alternately of strain and compression. The effective part of their force, in the present case, is not that which aids to *lift* or to *press* the superficial matter (for *that*, acting alike on the continents and on the bed of the sea, would have no influence), but that which tends to produce lateral displacement; or what geometers call the *tangential force.* This of necessity brings the whole ring of the earth's surface, which at any instant has the acting luminary *on its horizon*, into a state of strain; and the whole area over which it is nearly vertical, into one of compression. We leave this point to be further followed out, but we cannot forbear remarking, that the great volcanic chains of the world have, in point of fact, a direction which this cause of disruption would tend rather to favour than to contravene.

LECTURE II.

THE SUN.

THE subject I have chosen for this Lecture is perhaps an ambitious one; for it is no less than an attempt to convey to my hearers some faint impression of the vastness and grandeur of the most magnificent object in nature—of that glorious body which occupies the centre of our planetary system, and on which not only our own globe, but all the other planets, many of them of far greater magnitude and possibly too of greater importance in the scale of being than our own; depend in the most immediate manner for the fulfilment of those conditions without which animated existence and organic life are impossible—THE SUN. There is a poem of Byron's entitled "Darkness," which begins thus:—

> "I had a dream which was not all a dream,
> The bright sun was extinguish'd, and the earth
> Did wander darkling in th' eternal space
> Rayless and pathless,"

and so on: describing, or trying to describe, the horrors of that desolation which would ensue. They are assembled and piled on one another in this powerful poem with the hand of a master of the horrible; and in the end everybody goes mad, fights with everybody else, and dies of starvation.

(2.) But there would not be time for fighting or starvation. In three days from the extinction of the sun there would, in all probability, not be a vestige of animal or vegetable life on the globe; unless it were among deep-sea fishes and the subterranean inhabitants of the great limestone caves. The first forty-eight hours would suffice to precipitate every atom of moisture from the air in deluges of rain and piles of snow, and from that moment would set in a universal frost such as Siberia or the highest peak of the Himalayas never felt—a temperature of between two and three hundred degrees below the zero of our thermometers. This is no fanciful guess-work. Professor Tyndall has quite recently shown that it is entirely to the moisture existing in the air that our atmosphere owes its power of confining, and cherishing as it were the heat which is always endeavouring to radiate away from the earth's surface into space. Pure *air* is *perfectly* transparent to terrestrial heat—so that but for the moisture present in the atmosphere, every night would place the earth's surface as it were in contact with that intense cold which we are certain exists in empty space: a degree of cold which from several different and quite independent lines of inquiry we are sure is not less than 230 degrees of Fahrenheit's thermometer below the zero of that scale. No animal or vegetable could resist such a frost for an hour, any more than it could live for an hour in boiling water. Such a frost exists, no doubt, over the dark half of the moon, which has no atmosphere, neither air nor vapour, and in all probability quite as violent an extreme of heat, a boiling temperature at least, over the bright half; so that we may pretty well make up our minds as to that half of the moon at least which we see, being uninhabited; while on the other hand, if it would not lead too far away from our immediate subject, I think it might be shown on admissible principles, that Venus and Mercury, in spite of their nearness to the sun, and possibly also Jupiter and Saturn, in spite of their remoteness, *may* have climates in which animal and vegetable life such as we see them here, might be maintained.

(3.) But it is with the sun itself that we are now concerned. What I am going to say about the sun will consist of a series of statements so enormous in all their proportions, that I dare say, before I have done, some of my hearers will almost think me mad, or intending to palm on them a string of rhodomontades, like some of the mythical stories of the Hindus. And yet there is nothing more certain in modern science than the truth of some of the most extravagant of these statements; and, wild as they may seem to those who for the first time hear them, they appear not only not extravagant, but actually dwarfed into littleness by the still vaster revelations of that science respecting the scale of the visible universe; in every part of which when we come to measure in figures either the magnitude or the minuteness of its mechanisms, we find our arithmetic almost breaking down in the attempt, and numbers of ten or twenty places of figures, as it were tossed about like dust, and turning up on every occasion.

(4.) To come then to our subject. The first and most important office the sun has to perform in our system is to keep it together, to keep its members from parting company, from *seceding*, and running off into outer darkness, out of the reach of the genial influence of his beams. Were the sun simply *extinguished*, the planets would all continue to circulate round it as they do at present, only in cold and darkness; but were it annihilated, each would from that moment set forth on a journey into infinite space in the direction in which it happened then to be moving; and wander on, centuries after centuries, lost in that awful abyss which separates us from the stars, and without making any sensible approach even to the nearest of them in many hundreds or even thousands of years. The power by which the sun is enabled to perform this office—to gather the planets round its hearth and to keep them there—is the same in kind (though very different in intensity) with that which when a stone is thrown up into the air draws it down again to the earth. As to the manner in which this is effected by the weight of the stone, or its tendency to fall straight down, acting to turn or draw it out of its right-lined course oblique to the surface, and oblige it to move in a curve,—with the explanation of that we have here nothing to do.

That belongs to mechanics, and we must take it for granted. But in order to understand how it is possible to pass from this familiar case that we see every day before our eyes, to that of a vast globe like the earth revolving in an orbit about the sun, it will be necessary to enlarge the scale of our ideas of magnitude. We must try to conceive a similar degree of command and control exercised over such a mass as our globe, and over the much greater masses of the remote planets, by the sun as a central body; hardly moved from its place, while as it were swinging all the others round it. And for this purpose it is necessary to possess some distinct conception of what sort of a body the sun really is—of its size—of its distance from us—of its weight or mass—and of the proportion it bears to the other bodies, the earth included, which circulate round it.

(5.) It is strange what crude ideas people in general have about the size of very distant objects. I was reading only the other day a letter to the *Times* giving an account of a magnificent meteor. The writer described it as *round, about the size of a cricket-ball, and apparently about* 100 *yards off.* Many persons spoke of the tail of the great comet of 1858 as being several yards long, without at all seeming aware of the absurdity of such a way of talking. The sun or the moon may be covered by a three-penny piece held at arm's length: but it takes a house, or a church, or a great tree to cover it on a near horizon, and a hill or a mountain on a distant one; so that it must be at least as large as any of these objects. Among the ancient Greek philosophers there was a lively dispute as to the real size of the sun. One maintained that it was "precisely as large as it looks to be," a thoroughly Greek way of getting out of a difficulty. All the best thinkers among them, however, clearly saw that it must be a very large body. One of them (Anaxagoras) went the length of saying that it might be as large as all Greece, for which he got laughed at. But he was outbid by Anaximander, who said it was twenty-eight times as large the earth. What would Anaximander or the scoffer of Anaxagoras have said, could he have known what we now know, that, seen from the same distance as the sun, the territory of Greece would have been absolutely invisible; and that even the whole earth, if laid upon it, would not cover more than one thirteen-thousandth part of its apparent surface,—less in proportion, that is to say, than a single letter in the broad expanse of type which meets the reader's eye when a closely-printed volume with a large page and small type lies open before him.*

(6.) My object in this notice is not to put before my audience, except in one single instance, any connected chain of reasoning and deduction; or to show how, from the principles of abstract science combined with observation, the results I have to state have been obtained. This would lead me a great deal too far, and would require not one but a whole series of such lectures. What I

* The original type and page of "Good Words" were here referred to, in which this lecture first appeared in print: each page of which contains about 6000 letters. The pages which now lie open before the eye of the reader contain, together, only about 2600

aim at is to convey to their minds, as matters of fact, what those results *are* in the case of the sun, and to enable them to form a conception of it as a reality. Still it is reasonable for any one to ask how it is possible to prove such a statement, for instance, as that just made: and as the kind of process by which our conclusions as to the size and mass of the sun are arrived at may be put in a few words, it will not be amiss to give a sketch of it.

(7.) The first step towards ascertaining the real size of the sun is to determine its distance. Now, the simplest way to find the distance of an object which cannot be got at, is to measure what is called a base line from the two ends of which it can be seen at one and the same moment, and then to measure with proper instruments the angles at the base of the triangle formed by the distant object and the two ends of the base. Geography and surveying in modern times have arrived at such perfection, that we know the size and form of the earth we stand upon to an extreme nicety. It is a globe a little flattened in the direction of the poles,—the longer diameter, that across the equator, being 7925 miles and five furlongs, and the shorter, or polar axis, 7899 miles and one furlong; and in these measures it is pretty certain that there is not an error of a quarter of a mile. And knowing this, it is possible to calculate with quite as much exactness as if it could be measured, the distance *in a straight line* between any two places whose geographical positions on the earth's surface are known. Now there are two astronomical observatories very remote from one another; the one in the northern hemisphere, the other in the southern, viz., at Hammerfest in Norway, and at the Cape of Good Hope, both very nearly on the same meridian, so that the sun, or the moon, or any other heavenly body attains its greatest altitude above the horizon of each (or as astronomers express it, passes the meridian of each) very nearly at the same time. Supposing then that this, its *meridian altitude*, is carefully observed at each of these two stations on the same day; it is easy to find, by computation, the angles included between each of the two lines of direction in which it was seen from the two places, and their common line of junction; so that taking this latter line for the base of a triangle, of which the two sides are the distances of the object from either place, those two sides can thence be calculated by the very same process of computation which is employed in geographical surveying to find the distance of a signal from observations at the ends of a measured base. Now, the distance between Hammerfest and the Cape in a straight line is nearly 6300 miles, and owing to the situations of the two places in latitude, the triangle in question is always what a land surveyor would call a favourable one for calculation: so that, with so long a base, we may reasonably expect to arrive at a considerably exact knowledge of its sides,—after which a little additional calculation will readily enable us to conclude the distance of the object observed from the earth's centre.

(8.) When the moon is the object observed, this expectation is found to be justified. The triangle in question, though a long one, is not extravagantly so. Its sides are found to be, each about thirty-eight times the length of the base, and the resulting distance of the moon from the earth's centre about thirty diameters of the latter, or more exactly sixty times and a quarter its radius, that is to say, 238,100 (say 240,000) miles, which is rather under a quarter of a million—so that, speaking roughly, we may consider the moon's orbit round the earth as a circle about half a million of miles across. In the case of the sun, however, it is otherwise. The sides of our triangle *are* here what may be called extravagantly out of proportion to its base; and the result of the calculation is found to assign to the sun a distance very little short of four hundred times that already found for the moon—being in effect no less than 23,984 (in round numbers 24,000) radii, or 12,000 diameters of the earth, or in miles 94,880,700 or about 95,000,000.*

(9.) When so vast a disproportion exists between the distance of an object and the base employed to measure it, a very trifling error in the measured angles produces a great one in the result. Happily, however, there exists another and a very much more precise method, though far more refined in principle, by which this most important element can be determined; viz., by observations of the planet Venus, at the time of its "transit" (or visible passage) across the sun's disc. It would lead us too far aside from our purpose to explain this, however, at

* These numbers and all the subsequent statements in miles are too large by about 1 mile in 31. See Lecture III. on Comets, § 9.

length. The necessary observations were made at the time of the last "transit" in 1769, and will no doubt be repeated on the next occasion of the same kind, in 1874.*

(10.) From the distance of the sun so obtained, and from its apparent size (or, as astronomers call it, its angular diameter), measured very nicely by delicate instruments called micrometers, the real diameter of the sun has been calculated at 882,000 miles, which I suppose may be taken as exact to a few odd thousands.

(11.) Now, only let us pause a little, and consider among what sort of magnitudes we are landed. It runs glibly over the tongue to talk of a distance of 95,000,000 of miles, and a globe of 880,000 miles in diameter, but such numbers hardly convey any distinct notion to the mind. Let us see what kind of conception we can get of them in other ways. And first then, as to the distance. By railway, at an average rate of 40 miles an hour one might travel round the world in 26 days and nights. At the same rate it would take 270 years and more to get to the sun. The ball of an Armstrong 100-pounder leaves the gun with a speed of about 400 yards per second. Well, at the same rate of transit it would be more than thirteen years and a quarter in its journey to reach the sun; and the sound of the explosion (supposing it conveyed through the interval with the same speed that sound travels in our air), would not arrive till half a year later. The velocity of sound, or of any

* The distance above stated is that which results from this more precise mode of procedure. See this explained in Lecture V., § 17.

other impulse conveyed along a steel bar, is about sixteen times greater than in air. Now, suppose the sun and the earth connected by a steel bar. A blow struck at one end of the bar, or a pull applied to it, would not be delivered—would not begin to be felt—at the sun till after a lapse of 313 days. Even light, the speed of which is such that it would travel round the globe in less time than any bird takes to make a single stroke of his wing, requires seven minutes and a half to reach us from the sun.

(12.) The illustration of the distance of the sun which I have just mentioned, by supposing it connected with the earth by a steel bar, will serve to give us some notion of the wonderful connexion which that mystery of mysteries, gravitation, establishes between them. The sun *draws* or pulls the earth towards it. We know of no material way of communicating a pull to a distant object more immediate, more intimate, than grappling it with bonds of steel; and how such a bond would suffice we have just seen. But the *pull* on the earth which the sun makes is instantaneous, or at all events incomparably more rapid in its transmission across the interval than any solid connexion would produce, and even demonstrably far more rapid than the propagation of light itself.*

(13.) Let me now try to convey some sort of palpable notion of the *size* of the sun itself. On a circle six feet in diameter, representing a section of it through the centre, a similar section of the earth would be about

* See note at the end of this lecture.

represented by a fourpenny-piece, and a distance of a thousand miles by a line of less than one-twelfth of an inch in length. A circle concentric with it, representing on the same scale the size of the moon's orbit about the earth, would have for its diameter only thirty-nine inches and a quarter, or very little more than half the sun's. Imagine, now, if you can, a globe concentric with this earth on which we stand; large enough not only to fill the whole orbit of the moon, but to project beyond it on all sides into space almost as far again on the outside! A spangle, representing the moon, placed on the circumference of its orbit so represented, would require to be only a sixth part of an inch in diameter.

(14.) It is nothing to have the size of a giant without the strength of one. The sun retains the planets in their several orbits by a powerful mechanical force, precisely as the hand of a slinger retains the stone which he whirls round till the proper moment comes for letting it go. The stone pulls at the string one way, the controlling hand at the centre of its circle the other. Were the string too weak, it would break, and the stone, prematurely released, would fly off in a tangential direction. If a mechanist were told the weight of the stone (say a pound), the length of the string (say a yard, including the motion of the hand), and the number of turns made by the stone in a certain time (say sixty in a minute, or one in a second), he would be able to tell precisely what ought to be the strength of the string so as *just not to break;* that is to say, what weight it ought at least to be able to lift without breaking. In the case I have mentioned, it ought to be capable of sustaining 3 lb. 10 oz. 386 grs. If it be weaker it will break. And this is the force or effort which the hand must steadily exert, to draw the stone in towards itself, out of the direction in which it would naturally proceed if let go; and to keep it revolving in a circle at that distance.

(15.) Now, what the string does to the stone in the sling, that, in the case of the sun retaining the earth in its orbit, is done—that same office is performed—that *effort* (in some mysterious way which the human mind is utterly incapable of comprehending) is exerted—that pull communicated; in an instant of time, and so far as we can discover, without any material tie; by the force of gravitation. We know the time the earth takes to revolve about the sun. It is a year; of so many days, hours, minutes and seconds; and we know its distance —95,000,000 of miles, which may easily be turned into yards. Well, now, suppose a stone or a lump of lead of a ton weight to be tied to the sun by a string, and slung round it in such a circle and in such a time. Then, on the very same principles, and by the same rules of arithmetic, one may calculate the amount of pull, or tension of the string, and it will be found to come out 1 lb. 6 oz. 51 grs.

(16.) We all know what sort of lifting power—what amount of muscular force—it takes to sustain a pound weight. Multiply this by 2240 and you have the muscular effort necessary to sustain a ton. It would require three or four strong horses straining with all their might. Well, now, it is one of the peculiarities of this mysterious power of gravitation, that its intensity—the energy of its pull—is less and less as the distance of the thing pulled is greater: and *that* in a higher proportion. At double the distance, the force of the pull is not halved, but quartered: at triple, it is not a third part, but a ninth. There are mountains in the world five miles high; that is to say, whose summits are five miles farther from the centre of the earth than the sea-level. If a ton of lead were carried up to the top of such a mountain, though it would still balance another ton, or 2240 weights of a pound each on the scales, then and there; yet it would not require so great an effort, such an exertion of muscular force, to raise and sustain it by five pounds and a half. Now, fancy it removed to a height of 94,900,000 miles from the earth's surface, and estimating by the same rule its apparent weight, you will find, if you make the calculation, that it would not require more effort to sustain it from falling, than would suffice to lift one thirty-seventh part of a grain from the surface of the earth.

(17.) This, then, one thirty-seventh part of a grain, is the force which the earth, placed where the sun is, would exert on our lump of lead. But we have seen that to retain such a lump in such an orbit requires a pull of 1 lb. 6 oz. 51 grs. Of course, then, the earth, so placed, would be quite inadequate to retain it from flying off. To do this would require as many earths to pull it as there are thirty-seventh parts of a grain in 1 lb. 6 oz. 51 grs.: that is to say, by an easy sum in arithmetic, 356,929; or in round numbers, 360,000. Now, this is equivalent to saying, that to do the work which the sun does upon each individual ton of matter which the earth consists of, it must pull it as if (mind I say *as if*) it were made up of 360,000 earths. And this is what is meant by saying, that the mass or quantity of gravitating matter constituting the sun is 360,000 times as great as the mass or quantity of such matter in the earth.

(18.) Thus, now, you see, we have weighed as well as measured the sun, and the comparison of the two results leads to a very remarkable conclusion. In point of size, the globe of the sun, being *in diameter* 110 times that of the earth, occupies *in bulk* the cube of that number, or 1,331,000 times the amount of space. The disproportion in bulk, then, is much greater than the disproportion in weight,—very nearly four times greater: so that you see, comparatively speaking, and of course on an average of its whole mass, the sun consists of *much lighter* materials than the earth. And in this respect it agrees with all the four great exterior planets, Jupiter, Saturn, Uranus, and Neptune; while all the others—Mercury, Venus, and Mars—agree much more nearly with the earth, and seem to form a quite distinct and separate family.

(19.) From this calculation of the mass of the sun, and from its diameter, we are enabled to calculate the pressure which any heavy body placed on its surface would exercise upon it, or what power it would require to lift it off. It is very nearly thirty times the power required to lift the same mass here on earth. A pound of lead, for instance, transported to the sun's surface, could not be raised from it by an effort short of what would lift thirty pounds here. A man could no more stand upright there, than he could here on earth with twenty-nine men on his shoulders. He would be squeezed as flat as a pancake by his own weight.

(20.) Giant Size and Giant Strength are ugly qualities without beneficence. But the sun is the almoner of the Almighty, the delegated dispenser to us of light and warmth, as well as the centre of attraction; and as such, the immediate source of all our comforts, and indeed of the very possibility of our existence on earth. Even the very coals which we burn, owe their origin to the sun's influence, being all of vegetable materials, the remains of vast forests which have been buried and preserved in that form for the use of man, millions of ages before he was placed on the earth; and which, but for the solar light and heat, would have had no existence.* Indeed, the theory of heat which is now gaining ground would almost go to prove that it is the actual identical heat which the sun put into the coal, while in the form of living vegetation, that comes out again when it is burnt *as* coal in our grates and furnaces; so that, after all, Swift's idea of extracting sunbeams out of cucumbers, which he attributes to his Laputan philosophers, may not be so very absurd.†

* See the treatise on Astronomy, by the author of this paper, in "Lardner's Cabinet Cyclopædia," published in 1833. Stevenson (the celebrated engineer) has more recently drawn attention to this fact.

† Not more so at least than some of his other Laputan speculations; such as calcining ice into gunpowder: or moving vast locomotive masses by magnetism, both which feats have, in a somewhat altered form of expression, been accomplished (as in the explosion of potassium when laid on ice, and the movement of a ship by electro-mag-

(21.) But how shall I attempt to convey to you any conception of the scale on which the great work of warming and lighting is carried on in the sun? It is not by large words that it can be done. All "word-painting" must break down, and it is only by bringing before you the consideration of great facts in the simplest language, that there is any chance of doing it. In the very outset here is the greatest fact of all—the enormous waste, or what appears to us to be waste—the excessive, exorbitant prodigality of diffusion of the sun's light and heat. No doubt it is a great thing to light and warm the whole surface of our globe. Then look at such globes as Jupiter and Saturn and the others. This, as you will soon see, is something astounding; but then look what a trifling space they occupy in the whole sphere of diffusion around the sun. Conceive that little globe of the earth, such as we have described it in comparison with our six feet sphere, removed 12,000 of its own diameters, that is to say, 210 yards from the centre of such a sphere (for that would be the relative size of its orbit)! why, it would be an invisible point, and would require a strong telescope to be seen at all as a thing having size and shape. It occupies only the 75,000th part of the circumference of the circle which it describes about the sun. So that 75,000 of such earths at that distance, and in that circle placed side by side, would

netism); or than his plan for writing books by the concourse of accidental letters, and selection of such combinations as form syllables, words, sentences, &c., which has a close parallel in the learned theories of the production of the existing races of animals by natural selection.

all be equally well warmed and lighted,—and, then, that is only in one plane! But there is the whole sphere of space above and below, unoccupied; at any single point of which if an earth were placed at the same distance, it would receive the same amount of light and heat. Take all the planets together, great and small; the light and heat they receive is only one 227 millionth part of the whole quantity thrown out by the sun. All the rest escapes into free space, and is lost among the stars; or does there some other work that we know nothing about. Of the small fraction thus utilized in our system, the earth takes for its share only one 10th part, or less than one 2000 millionth part of the whole supply.

(22.) Now, then, bearing in mind this huge preliminary fact to start with, let us see what amount of *heat* the earth *does* receive from the sun. The earth is a globe; and therefore, taken on an average, it is constantly receiving as much, both of light and heat, as a flat circle 8000 miles in diameter, held perpendicularly to receive it. Now, that section is 50,000,000 square miles, so that there falls at every instant on the whole earth 50,000,000 times as much heat as falls on a square mile of the hottest desert under the equator at noonday with a vertical sun and with not a cloud in the sky—and in fact nearly a third more; for more than a quarter of the sun's heat is absorbed in the air in the clearest weather, and never reaches the ground. Now, we all know that in those countries it is much hotter than we like to keep our rooms by fires. I have seen the thermometer four inches deep in the sand in South Africa rise to 159° Fahrenheit:

and I have cooked a beef-steak and boiled eggs hard by simple exposure to the sun in a box covered with a pane of window-glass, and placed in another box so covered.

(23.) From a series of experiments I made there, I ascertained that the direct heat of the sun, received on a a surface capable of absorbing and retaining it, is competent to melt an inch in thickness of ice in 2 13^{m}, and from this I was enabled to calculate how much ice would be melted per hour by the heat actually thrown on a square mile exposed at noon under the equator, and the result is 58,360,000 lb., or in round numbers, 26,000 tons, and this vast mass, has to be multiplied 50 million-fold to give the effect produced on a diametral section of our globe.

(24.) And, now, let us endeavour to form some kind of estimate of the *temperature;* that is to say, the degree or intensity of the heat at the actual surface of the sun. By a calculation, with which I will not trouble you, it turns out to be more than 90,000 times greater than the intensity of sunshine here on our globe at noon and under the equator—a far greater heat than can be produced in the focus of any burning-glass; though some have been made powerful enough to melt, not only silver and gold, but even platina, and, indeed, all metals which resist the greatest heats that can be raised in furnaces.

(25.) Perhaps the best way to convey some sort of conception of it, will be to state the result of certain experiments and calculations recently published; which is this—that the heat thrown out FROM EVERY SQUARE YARD of the sun's surface is equal to that which would be produced by burning on that square yard six tons of coal per hour, and keeping up constantly to that rate of consumption—which, if used to the greatest advantage, would keep a 63,000 horse steam-engine at work.—And this, mind, on each individual square yard of that enormous surface which is 12,000 times that of the whole surface of the earth!

(26.) Let me say something now of the *light* of the sun. The means we have of measuring the intensity of light are not nearly so exact as in the case of heat—but this at least we know—that the most intense lights we can produce artificially, are as nothing compared *surface for surface* with the sun.—The most brilliant and beautiful light which can be artificially produced is that of a ball of quicklime kept violently hot by a flame of mixed ignited oxygen and hydrogen gases playing on its surface. Such a ball, if brought near enough to appear of the same size as the sun does, can no more be looked at without hurt than the sun—but if it be held between the eye and the sun, and *both* so enfeebled by a dark glass as to allow of their being looked at together—it appears as a black spot on the sun or as the black outline of the moon in an eclipse, seen thrown upon it. It has been ascertained by experiments which I cannot now describe, that the brightness, the intrinsic splendour, of the surface of such a lime-ball is only one 146th part of that of the sun's surface. That is to say, that the sun gives out as much light as 146 balls of quicklime *each the size of the sun*, and each heated *over all its surface* in the way I have described, which is the most intense heat we can raise, and in which platina melts like lead.

(27.) On the benefits which the sun's light confers on us it cannot be necessary to say much; only one thing, I think, may not be known to all who may read these pages, viz., that it is not only by enabling us to see that it is useful, but that it is quite as necessary as its heat to the life and well-being both of plants and animals. Animals, indeed, may live some time in complete darkness, but they grow unhealthy; lose strength and pine away; while plants very quickly lose their green colour; turn white or pale yellow; lose all their peculiar scent and flavour; refuse to flower; and at last rot and die off. What I have now to say about the light of the sun is of quite a different nature.

(28.) The sun's light, as we all know, is purely *white*. If the sun sometimes looks yellow or red, it is because it is seen through vapours, or smoke, or a London fog of smoke *and* vapour *mixed*. It has been seen blue;* but when high up, in a clear sky, it is quite white. The whiteness of snow, of a white cloud, of white paper, is the whiteness of the sun's light which falls upon them. Whatever reflects the rays of the sun *without choice or preference, appears white*. Whatever does not do so appears coloured; and if it does not reflect them at all—black. Now I must explain what I mean by saying—" without choice or pre-

* This has been denied by Arago. But I have a description of the phænomenon by an eye-witness, accompanied with a coloured drawing, which leaves no doubt on my mind of the reality of the fact. It was after a hurricane at Barbadoes.

ference." Every ray of light which comes from the sun is not a simple but a compound *thing*. Here, again, I must explain. The air we breathe is not a simple but a compound *thing*. It is separable at least into four distinct *things*, as different from one another as any four things you can name. Well, then, so of a ray or beam of the sun; it may be separated, split, subdivided, not into four, but into many hundreds, nay, thousands, of perfectly distinct rays or things, or rather of three distinct sorts or species of rays; of which one sort affects the eyes as light; one the sense of feeling and the thermometer as heat; and one the chemical composition of everything it falls upon; and which produces all the effects of photography. Each of these three classes (and I believe there are several more, indeed I have proved the existence of one more) consists of absolutely innumerable *species* or sorts; every one of which is separated from every other by a boundary line, as sharp and as distinct as that which separates Kent and Sussex on a map. A ray of light is a world in miniature, and if I were to set down all that experiment has revealed to us of its nature and constitution, it would take more volumes than there are pages in the manuscript of this lecture.

(29.) When the sun's light is allowed to pass through a small hole in a dark place, the course of the ray or sunbeam may be traced through the air (by reason of the small fine dust that is always floating in it), as a straight line or thread of light of the same apparent size, or very nearly so, from the hole to the opposite wall. But if in the course of such a beam, be held at any point the edge of a clear angular polished piece of glass called a *prism*, the course of the beam from that place will be seen to be bent aside in a direction towards the thicker part of the glass—and not only so bent or *refracted*, but spread out to a certain degree, so that the beam in its further progress grows continually broader, the light being *dispersed*, into a flat fan-shaped plane: and if this be received on white paper; instead of a single white spot which the unbroken beam would have formed on it, appears a coloured streak; the colours being of exceeding vividness and brilliancy, and following one another in a certain fixed order—graduating from a pure crimson red at the end least remote from the original direction (or least *deviated*), through orange, yellow, green, and blue, to a faint and rather rosy violet. This beautiful phænomenon—the *Prismatic Spectrum*, as it is called—strikes every one who sees it for the first time in a high degree of purity, with wonder and delight; as I once had the gratification of witnessing in the case of that eminent artist the late Sir David Wilkie, who, strange to say, had never seen a "Spectrum" till I had the pleasure of showing him one; and whose exclamations, though a man habitually of few words, I shall not easily forget. I shall not attempt to give any account of the theory of this *prismatic dispersion* of the sunbeam; but an illustration of it may be found in a very familiar and primitive operation—the winnowing of wheat. Suppose I had a sieve full of mixed grains and other things—shot, for instance; wheat grains; sand; chaff; feathers; and that I flung them all out across a side wind, and noticed where they fell. The shot would fall in one place, the wheat in another, the sand in another, the chaff in another, and the feathers anywhere—nowhere; but none of them in the straight direction in which they were originally tossed. All would be *deviated;* and if you marked the places of each sort, you would find them all arranged in a certain order—that of their relative lightness—in a line on the ground, oblique to the line of their projection. You would have separated and assorted them, and formed a *spectrum*, so to speak, on the ground; or a picture of what had taken place in the process; which would in effect have been the performance of a mechanical analysis of the contents of your basket.

(30.) Bearing always in mind that it is an illustration of a series of facts, not a theoretical explanation of a natural process, which is here intended; I will now proceed to observe that the analogy of this case to that of the prismatic analysis of a sunbeam may be pursued still further. If the original contents of the basket had been all of one material, such as sand, consisting of a mixture of particles of every gradation of coarseness and fineness; from small pebbles down to impalpable dust; the trace upon the ground, the sand spectrum, however long, would be uninterrupted: the coarsest particles lying at one end; the finest at the other; and every intermediate size in every intermediate place. On the other hand, in the case first supposed, and supposing the shot to differ *inter se* in respect of size within certain limits; the wheat grains again within certain other; the sand within other;

and so on; they would be found after projection all indeed lying in a line, but that line an interrupted one—consisting first of shot occupying a certain length; then an interval; then wheaten grains to a certain extent—another interval—then sand, chaff, and so on. Now this is by no means an inapt though a coarse representation of the constitution of the Prismatic Spectrum. When it is formed by an extremely pure prism, and with certain precautions (which need not here be detailed) to ensure the perfect purity of its colours, it is found to be *discontinuous:* that is to say, not a simple streak like a riband of paper coloured from end to end by tints graduating insensibly from red to violet, but like such a riband marked, *across its* breadth, by perfectly black lines of exceeding delicacy, yet some wider some narrower than others; and where these lines are, the paper is not illuminated at all. Into these spaces (for narrow as they are, they have each a certain breadth) none of the light dispersed by the prism falls. These lines, be it also observed, are not occasional or accidental, but permanent; and belong to the sun's light *as such.* They divide the spectrum into compartments as the boundary lines between counties on a map divide the soil into regions; and each individual of these compartments differs in other qualities besides colour from its neighbours on either side; much as contiguous regions of a country differ in soil and cultivation as well as in climate. It is as if our assorted grains were distinguished not only by being coloured according to their respective sizes, but each particular size and weight distinguished also by differences in the material of which they consisted.

(31.) Every observer who has examined the spectrum with more care than the last, has added to the number of these lines. Dr Wollaston first noticed two or three of the most conspicuous. Fraunhofer registered and fixed the places of some thirty or forty more; and later observers have mapped down with all the precision of a geographical survey, not less than two thousand of them. The knowledge of them, and the precise measurement of their distances from one another, has proved most valuable in a great many lines of scientific enquiry, and most particularly in Optics and Chemistry; and, quite recently, has been the means of revealing facts respecting the constitution of the sun itself, which one would have supposed it impossible for man ever to have become acquainted with. One word more on these lines—for we must husband time, as there remains a great deal more ground to go over. I have said that they are not *occasional*, but belong to the sun's light *as such*. But they may be considered as in some sort *accidental* as regards the sun—for the light of each of the stars when thrown into a spectrum, is found to have a different system of these "fixed lines." And what is more, the light of every flame has its peculiar lines, which indicate the nature of the burning substance. And in this way there seems to arise a possibility that by studying these lines carefully, as ex hibited by terrestrial flames and other sources of artificial light, we may come to a knowlege of what the sun and stars are made of. This is what men of science are now very busily occupied about, and it seems to have been rendered at least highly probable—I do not say that it has been proved—that a great many of the chemical elements of this our earth exist in the sun—such as, for instance, iron, soda, magnesia, and some others. We cannot here state the extraordinary facts on which this conclusion rests. But the conclusion itself is not so absolutely strange and startling as it may at first appear. The analysis of meteorolites, which there can be no doubt have come to the earth from very remote regions of the Planetary spaces, has, up to the present time, exhibited no new chemical element—so that a community of nature, at least as regards material constitution, between our earth and the rest of the bodies of our system, is at all events no unexpected, as it is, in itself, no unreasonable conclusion.

(32.) Not that it is meant, by anything above said, to imply that the light of the sun is that of any flame, in the usual sense of the word. A late celebrated French philosopher, M. Arago, indeed, considered that he had proved it to be so by certain optical tests. But in the first place his proof is vitiated by an enormous oversight; and the thing, besides, is a physical impossibility. The light and heat of the sun cannot possibly arise from the burning of *fuel*, so as to give out what we call flame. If it be the sun's substance that *burns* (I mean consumes), where is the oxygen to come from? and what is to become of the ashes, and other products of combustion? Even supposing the oxygen supplied from the material, as in the cases of gunpowder, Bengal light, or gun cotton, still the chemical products have to be disposed of. In the case of gun cotton, it has been calculated that, if the sun were made of it so condensed as only to burn on the surface, it would burn out, at the rate of the sun's expenditure of light and heat, in eight thousand years. Anyhow—fire, kept up by fuel and air, is out of the question. There remain only three possible sources of them, so far as we can perceive—electricity, friction, and vital action. The first of these was suggested by the late Sir William Herschel in 1801; the second, at least as a possibility, though without indicating any mode by which the necessary friction could arise, by myself, in a work* published in 1833. The theory at present current of it is founded on what may not unfairly be considered a further development of this idea, the friction being supposed to arise from meteoric matter circulating round the sun, and gradually subsiding into it, and either tearing up its surface, or ploughing into its atmosphere. But on this we cannot dilate, as nothing has been hitherto said about the appearance of the sun in telescopes, and the strange phænomena its surface, so examined, exhibits.

(33.) One of the earliest applications of the telescope was to turn it on the sun. And the first fruits of this application (which originated about the same time in the year 1611, with Harriot in England, Galileo in Italy, and Fabricius and Scheiner in Germany), was the dis-

* "Lardner's Cabinet Cyclopædia," Astronomy, s. 337, p. 212. Aristotle was earlier in making this suggestion: but such random guesses as those of the ancients can hardly merit the name of scientific suggestions.

covery of black spots on its surface, which, when watched from day to day, were found to change their situation on its disc, in a certain regular manner; coming in, or making their first appearance on the eastern edge or border of the disc: *i.e.*, on the left-hand side of the sun when seen at noonday; and going off, or disappearing at the west, or on the right-hand side. It very soon became evident that, whatever these spots might be, they adhered to the body of the sun, and that their apparent motions could only be accounted for by a real motion of rotation of the sun on an axis nearly, but not quite, perpendicular to the ecliptic. By following out this indication by careful observation and calculation, it has become known that the sun does so rotate; that the time occupied in a single rotation is very nearly 25 days 7 hours 48 minutes; that the axis of rotation is about 7° inclined to a line perpendicular to the ecliptic, its direction in space being that of a line pointing nearly to the star τ (tau), in the constellation of the Dragon; in consequence of which on and about the 11th of June, the spots appear to pass across the sun in straight lines, from the apparent northern to the apparent southern hemisphere of the sun, and the reverse on and about the 12th of December, while at intervening times, their course across the sun is a flattened elliptical or oval curve; a necessary consequence of their real motion being in a circle much inclined to the line of sight. Their ellipses are most open on the 11th of March, and the 13th of September; on the former of which days we get the best view of the south pole of the sun, and on the latter of the north.

(34.) But here comes the strange part of their history. These spots are not permanent marks on the sun's surface. They come and go. They begin as small dim specks; grow to be great blotches; and then dwindle away. Sometimes they are large enough to be seen without a telescope, when the sun is near setting or just risen, so as to have its dazzling splendour mitigated by the vapours of the horizon, and admit of being looked at steadily. Many instances of such appearances are recorded, some very remarkable ones, long before the invention of the telescope. Two were so seen by my son, Mr A. Herschel, in London, in November, 1861, who sent me a drawing of them, which I found verified on comparison with a drawing taken from the telescope on the same day, by a very assiduous observer in my immediate neighbourhood.

(35.) Ever since the first discovery of the solar spots, they have been watched with great interest, and it has been ascertained that they do not make their appearance indiscriminately upon every part of the globe of the sun. At or near either of its poles they never appear; and very rarely indeed on its equator, or on any part of its body beyond the 40th degree of latitude—understanding that term on the sun in the same acceptation which geographers attach to it on our own globe. They mainly frequent two zones or belts parallel to its equator; bearing very nearly the same relation to that great circle of its sphere which the regions on our own globe in which the trade winds prevail, bear to the equatorial region of the earth—extending, that is to say, to some

25° or 30° of north, and not quite so far, or in such abundance in south latitude; with a *comparatively* spotless intermediate belt, of five or six degrees broad between them, answering to our region of equatorial calms. The resemblance is so striking as most strongly to suggest some analogy in the causes of the two phænomena—and it has been suggested that as our trade winds originate in a *greater influx* of heat from without, on and near the equator, than at the poles, combined with the earth's rotation on its axis: so the maculiferous belts of the sun may owe their origin to a *less* * equatorial *efflux* of heat, combined with the axial rotation of that luminary.† There is another extremely remarkable feature in the appearance and disappearance of these spots. I have said that they are not *permanent*. Sometimes, indeed, but rarely, one and the same spot lasts long enough, after disappearing at the western edge of the sun, to come round again and reappear at the eastern; and it has happened that a spot has lasted long enough to reappear four or five times; but for the most part this is not the case. But as regards the number of spots which appear on the sun at different times, there is the greatest possible difference. Sometimes it is quite spotless; at others the spots swarm upon it: and as many as fifty or sixty spots or groups, large and small, have been seen at once, arranged in two belts.

(36.) Now, it has lately been ascertained by a careful

* Misprinted *greater* in the original lecture as it appeared in *Good Words*.

† "Results of Astronomical Observations at the Cape of Good Hope," by the author, p. 434.

comparison of all the recorded observations of the spots, that the periods of their scarcity and abundance succeed one another at regular intervals of a trifle more than five years and a half: so that in eleven years and one-tenth, or nine times in a century, the sun passes through all its states of purity and spottiness. Thus, for instance, in the present century, the years 1800, 1811, 1822, 1833, 1844, 1855–6 were years in which the sun exhibited few or no spots, while in the years 1805, 1816, 1827, 1838, 1849, 1860, the spots have been remarkably abundant and large. Several attempts have been made to connect this with periodical variations in the weather, with hot and cold years—wet and dry ones—years of good and bad harvests, etc.; but though I believe there is some such connexion, it is so overlaid and, as it were, masked by the multitude of causes which act to produce what we call the prevalent weather of a season, that nothing satisfactory has been made out. But there are two classes of phænomena or facts which occur here on earth which certainly do stand in very singular accordance with the appearance and disappearance of the sun's spots. The first is that splendid and beautiful appearance in the sky which we call the Aurora or Northern lights; and which, by comparison of the recorded displays, have been ascertained to be much more frequent in the years when the spots are abundant, and extremely rare in those years when the sun is free from spots. The other is a class of facts not so obvious to common observation, but of very great importance to us; because it is connected with the history and theory of the mariner's compass, and with the magnetism of the earth; which we all know to be the cause of the compass needle pointing to the north. This is only a rough way of speaking. It does *not* point to the north, but very considerably to the west of north, and that, not always alike. Three centuries ago it pointed nearly as much east as now west of north. From year to year the change is very perceptible; and, what is more, at every hour of the day there is a small but perfectly distinct movement to and fro, eastward and westward, of its average direction. But besides this, the compass needle is subject to irregular, sudden, and capricious variations—jerking, as it were, aside, and oscillating backwards and forwards without any visible cause of disturbance. And, what is still more strange; these disturbances and jerks sometimes go on for many hours and even days, and often at the same instants of time, over very large regions of the globe; and in some remarkable instances, over the whole earth—the same jerks and jumps occurring at the same moments of time (allowance made for the difference of longitude). These occurrences are called magnetic storms, and they invariably accompany great displays of the Aurora; and are very much more frequent when the sun is most spotted, and rarely or never witnessed in the years of few spots.

(37.) The last four years* have been remarkable for spots, and there occurred on the 1st September 1859, an appearance on the sun which may be considered an epoch, if not in the sun's history, at least in our know-

* This lecture was delivered about the end of 1861.

ledge of it. On that day great spots were exhibited; and two observers, far apart and unknown to each other, were viewing them with powerful telescopes; when suddenly, at the same moment of time, both saw a strikingly brilliant luminous appearance, like a cloud of light far brighter than the general surface of the sun, break out in the immediate neighbourhood of one of the spots, and sweep *across* and *beside* it. It occupied about five minutes in its passage, and in that time travelled over a space on the sun's surface which could not be estimated at less than 35,000 miles.

(38.) A magnetic storm was in progress at the time. From the 28th of August to the 4th of September many indications showed the earth to have been in a perfect convulsion of electro-magnetism. When one of the observers I have mentioned had registered his observation; he bethought himself of sending to Kew, where there are self-registering magnetic instruments always at work, recording by photography at every instant of the twenty-four hours the positions of three magnetic needles differently arranged. On examining the record for that day, it was found that at that very moment of time (as if the influence had arrived with the light) all three had made a strongly marked jerk from their former positions. By degrees, accounts began to pour in of great Auroras seen on the nights of those days; not only in these latitudes, but at Rome; in the West Indies; on the tropics within 18° of the equator (where they hardly ever appear), nay, what is still more striking, in South America and in Australia; where, at Melbourne, on the night of the 2d of September the greatest Aurora ever seen there made its appearance. These Auroras were accompanied with unusually great electro-magnetic disturbances in every part of the world. In many places the telegraphic wires struck work. They had too many private messages of their own to convey. At Washington and Philadelphia, in America, the telegraph signal-men received severe electric shocks. At a station in Norway the telegraphic apparatus was set fire to; and at Boston, in North America, a flame of fire followed the pen of Bain's electric telegraph, which, as my hearers perhaps know, writes down the message upon chemically prepared paper.

(39.) I must now proceed to tell you what the telescope has revealed to us as to the nature and magnitude of these spots. And here again, the closer we look, the more the wonder increases. The spots were at first supposed to be clouds of black smoke floating over the great fiery furnace beneath,—then great lumps of fresh coal laid on; then comets fallen in to feed the fire; then tops of mountains standing up above a great surging ocean of melted matter. They are none of all these things; they are not clouds floating above the light, nor protuberances sticking up above the general surface; they are regions in which, by the action of some most violent cause, the bright, luminous clouds, or what at all events we may provisionally call clouds, which float in the sun's atmosphere, are for a time cleared off; and through the irregular vacuities thus created, allow us to see perhaps thousands or tens of thousands of miles below them, *first*, a layer of what we may consider real clouds, which appear comparatively dark, *as if* they were not self-luminous, but were seen only by the reflected light of the upper layer of bright ones; *secondly*, through other openings in this first layer, a second still darker layer, independent of the first, and probably still thousands of miles below that, and reached by some—but very little — light from above; and *thirdly*, through again other openings, what at present we must consider to be the body of the sun itself—at some vast and immeasurable depth still lower—and emitting so little light in comparison as to appear quite black, though that does not prevent its being in as vivid a state of fiery glare as a white-hot iron; when we remember what has been said of the lime light appearing black against the light of the sun's surface. And it is a fact, that when Venus, and Mercury pass across the sun, and are seen as round spots on it, they do really appear sensibly blacker than the blackest parts of the spots.

(40.) The sun then has an atmosphere, and in that atmosphere float at least three layers of something, that, for want of a better word, we must call clouds. The two nearest the body are not luminous. They cannot possibly be clouds of *watery* vapour, such as we have in our air, for water in a non-transparent state could not exist at that heat; but they may be what perhaps we might call smokes, that is to say, clouds in which the metals or their oxides and the earths exist in the same intermediate form that water does in our clouds. The third or upper layer of luminous clouds, or, as it is called, "the photosphere," is a sort of thing that three or four years ago we might be said to know nothing at all about; I mean as to its nature and constitution; but within that time a most wonderful discovery has been made by Mr Nasmyth. According to his observations, made with a very fine telescope of his own making, the bright surface of the sun consists of separate, insulated, individual objects or *things*, all nearly or exactly of one certain definite size and shape, which is more like that of a willow leaf, as he describes them, than anything else. These leaves or scales are not arranged in any order (as those on a butterfly's wing are), but lie crossing one another in all directions, like what are called spills in the game of spillikins; except at the borders of a spot, where they point for the most part inwards towards the middle of the spot, presenting much the sort of appearance that the small leaves of some water-plants or sea-weeds do at the edge of a deep hole of clear water. The exceedingly definite shape of these objects; their exact similarity one to another; and the way in which they lie across and athwart each other (except where they form a sort of bridge across a spot, in which case they seem to affect a common direction, that, namely, of the bridge itself),—all these characters seem quite repugnant to the notion of their being of a vaporous, a cloudy, or a fluid nature. Nothing remains but to consider them as separate and independent sheets, flakes, or scales, having some sort of solidity. And these flakes, be they what they may, and whatever may be said about the dashing of meteoric stones into the sun's atmosphere, etc., are evidently the *immediate sources of the solar light and heat,* by whatever mechanism or whatever processes they may be enabled to develop and, as it were, elaborate these elements from the bosom of the non-luminous fluid in which they appear to float. Looked at in this point of view, we cannot refuse to regard them as *organisms* of some peculiar and amazing kind; and though it would be too daring to speak of such organization as partaking of the nature of life, yet we do know that vital action is competent to develop both heat, light, and electricity. These wonderful objects have been seen by others as well as by Mr Nasmyth, so that there is no room to doubt of their reality. To be seen at all, however, even with the highest magnifying powers our telescopes will bear when applied to the sun, they can hardly be less than a thousand miles in length, and two or three hundred in breadth.

(41.) Next as to the actual size of the spots themselves: the distance of the sun is so vast, that a single second of angular measure on its surface as seen from the earth corresponds to 460 miles; and since, to present a distinguishable *form*, so as to allow of a certainty, for instance, that it is round or square, in the best telescopes, an object must present a surface of at least a second in diameter, it follows that to be seen at all so as to make out its shape, a spot must cover an area of not less than two hundred thousand square miles. Now, spots of not very irregular, and what may be called a compact form, of two minutes in extent, covering, that is to say, an area of between seven and eight hundred millions of square miles, are by no means uncommon. One spot which I measured in the year 1837 occupied no less than three thousand seven hundred and eighty millions, taking in all the irregularities of its form; and the black space or "umbra" in the middle of one, which was very nearly round, would have allowed the earth to drop through it, leaving a thousand miles clear of contact on every side: and many instances of much larger spots than these are on record. What are we to think, then, of the awful scale of hurricane and turmoil and fiery tempest which can in a few days totally change the form of such a region, break it up into distinct parts—open up great abysses in one part, such as that I have just described, and fill up others beside them? As to the forms of the spots, they are so conspicuously irregular as to defy description.

(42.) But we must proceed, for there are more wonders yet to relate. Far beyond the photosphere, or brilliant surface of the sun, extends what perhaps may be considered as its true atmosphere. This can only be seen at all in the rare opportunities afforded by total eclipses of the sun. Everybody knows that an eclipse of the sun is caused by the moon coming between it and us. Now, by an odd coincidence, it so happens that the sun being 400 times farther off than the moon, is also ALMOST exactly, but a trifle less than 400 times as large in *diameter;* so that when the centre of the moon comes exactly in the line with the centre of the sun it appears to cover it, and a very little more, so as to project on all sides a very little beyond it. Now, as the moon is opaque (or not transparent), it completely stops all the light from every part of the bright disc of the sun, so long as the *total* eclipse continues, which is sometimes as much as two or three minutes; and then are witnessed, what at no other time can be seen, viz., certain wonderful appearances of rose-coloured masses of light projecting, as it were, from the dark edge of the moon, for the most part like knobs, or cones, or long ranged ridges of what would seem to be mountains, rising from it; but sometimes like clouds or flaring flag-shaped masses of red light, some of which have been seen quite detatched from all connexion with the moon's border. That they belong to the sun, however, and not the moon, is evident from the fact that the moon in its progress over the sun's face gradually *hides* those *to* which it is approaching, and discloses those which belong to that side of the sun which the moon is going to leave; for I should mention that they are seen irregularly placed all round the edge of the sun.

(43.) Now, what *are* these singular lights? Flames they certainly are not; clouds of some sort it is extremely probable that they are, of most excessively thin and filmy vapour, floating in a transparent atmosphere which must for that purpose extend to a very considerable height above the luminous surface of the sun. We are all familiar with the beautiful appearance of those thin vapoury clouds which appear in our own atmosphere at sunset. But these solar clouds must be almost infinitely thinner and more unsubstantial, since even in that intense illumination they are only seen when the sun itself is hidden; and when it is remembered that the head of the comet of 1843 was seen at noon-day within two or three degrees of the sun by the naked eye.

(44.) Then, again, as to the magnitude of these cloudy masses, it must be enormous. Some of them have projected or stood out from the edge of the sun to a distance calculated at no less than forty or fifty thousand miles. They have now been observed in three great eclipses, that of 1842, 1851, and 1859; on which last occasion they were photographed in Spain by Mr De la Rue, under such circumstances as left no possibility of doubting their belonging to the sun. I dwell upon this, because there is another luminous appearance seen about the moon in total eclipses of the sun, which can only be referred to vapours of excessive tenuity, existing at an immense height in our own atmosphere; and which surrounds the disc of the moon like a glory, or *corona*, as it is called. By the accounts of all who have witnessed a total eclipse of the sun, it is one of the most awful natural phænomena. An earthquake has "rolled unheededly away" during a battle, but an eclipse has on more than one occasion either stopped the combat or so paralyzed one of the parties with terror, as to give the others who were prepared for it an easy victory: and I may as well add that two very remarkable battles in ancient history, the one on the 28th May, B.C. 585, the other the 19th May, B.C. 557, which were in progress during total eclipses, have had the years and days of their occurrence thereby fixed by calculation with a certainty which belongs to no other epochs in ancient chronology.

(45.) There is only one more point which my limits will allow me to touch upon. I will go back to my original metaphor. Our giant may be a huge giant and a strong giant, and a good-natured giant, but if he be a sluggard he is no giant worth the name. We have seen that he is a little slow to turn on his axis and roll himself round in his nest. But take him in his relation to the outer world, he is lively enough; he "rejoices *as* a giant to run his course;" and vindicates his credit as a swift runner with a vengeance! Hitherto I have only spoken of the sun *as* a sun, the centre of our system; and, as such, regarded by us as immovable. Even in this capacity he is not *quite* fixed. If he pulls the planets, they pull him and each other; but such family struggles affect him but little. *They amuse them*, and set them dancing rather oddly; *but don't disturb him.* As all the gods in the ancient mythology hung dangling from and tugging at the golden chain which linked them to the throne of Jove; but without power to draw him from his seat: so if all the planets were in one straight line, and exerting their joint attractions, the sun, leaning a little back as it were to resist their force, would not be displaced by a space equal to his own radius; and the fixed centre, or, as an engineer would call it, the centre of gravity of our system, would still lie within the sun's globe.

(46.) But the sun has another and, so far as we can judge, a much vaster part in creation to perform than to sit still as the quiet patriarch of a domestic circle. He is up and active as a member of a community like himself. The sun is not only a *sun*, he is a STAR also, and that but a small one in comparison with individual stars (one of which, Sirius, would make two or three hundred of him); and among these glorious compeers he moves on a path which is just beginning to become known to us; though in what orbit, or for what purpose, will never be given to man to know. Yet we do know—almost to a nicety—the direction in which that path is leading; and the rate of his travel (though this is less exactly determined). Still this rate, at the very lowest estimate, cannot be taken under four or five hundred thousand miles a day; and yet this speed, vast as it is, in the 2000 years which separate us from the observations of Hipparchus (who made the first catalogue of the stars), would not suffice to carry it (and of course our system along with it) over one sixtieth part of the distance which now separates it from the very nearest of the stars. When we travel through a diversified country, we become aware of our change of situation by the different grouping and presentation of the objects around us. But though travelling at this amazing rate through space, successive generations of mankind witness no change in the order and arrangement of the stars; and Hipparchus, were he to come once more among us, would recognize the old familiar forms of his constellations; and, without better means of observation than he then possessed, would be unable to detect, with certainty, any change in their appearance; though we, who are better provided in that respect, are enabled to do so.

(47.) Such, then, is the scale of things with which we

become familiar when we contemplate the sun. In what has been said, it will be perceived that I have been more anxious to dwell upon facts than theories, and rather to supply the imaginations of my audience with materials for forming a just conception of the stupendous magnificence of this member of God's creation, than to puzzle them with physical and mathematical reasonings and arguments.

NOTE ON § 12.—The effect of any supposed small loss of time in the transmission of the sun's attractive force on the earth across the intervening space, may be very easily made intelligible without going through any abstruse calculation. The *pull* exerted on the earth would be *delivered* there, not in the direction of the line joining the sun and earth at the instant of its arrival, but of that which *did* join them when it left the sun. Its action on the earth would therefore be oblique to their actual line of junction, or to what is called the radius vector of the orbit—tending, not *towards the sun*, but towards a point somewhat in advance of it—(*i.e.*, lying from it in the direction in space of the region towards which the earth is moving). This force then being resolved in radial and tangential directions would produce, in the former, a force directed to the sun differing by a mere infinitesimal from its direct gravity—and in the latter, one *always accelerating the earth in its orbit*, and which, however minute, must of necessity result in a continually progressive increase of the major axis, and therefore of the length of the year. Supposing the transmission of gravity to be performed with the speed only of light —the inclination of the line of pull to the radius vector would be $20''\cdot25$ (the exact value of the coefficient of aberration), and the accelerating tangential force thence resulting would amount to 1-10188th part of the sun's direct attraction, a force whose effects would become evident in a very few years—to say nothing of the centuries elapsed since the first determination of the length of the year.

LECTURE III.

ON COMETS.

THE subject of comets, about which I now propose to say something, is one that has of late naturally drawn to it a good deal of inquiry and general interest, by reason of the unusually magnificent spectacles of this description which have within the last few years been exhibited to us.* In itself it is perhaps not one of the best adapted for popular discussion and familiar explanation of this nature, because there are so many things in the history of comets unexplained, and so many wild and extravagant notions in consequence floating about in the minds of even well-informed persons, that the whole subject has rather, in the public mind, that kind of dreamy indefinite interest that attaches to signs and wonders than any distinct, positive, practical bearing. The fact is, that, though much is certainly known about comets, there

* This lecture was delivered on February 14, 1859.

is a great deal more about which our theories are quite at fault; and, in short, that it is a subject rather calculated to show us the extent of our ignorance than to make us vain of our knowledge, and to cause us to exclaim with Hamlet, "There are more things in heaven and earth, Horatio, than are dreamt of in our philosophy." This; the sublimity of the spectacle they afford; and the universal interest they inspire, make the appearance of a great comet an occasion for the imaginations of men to break loose from all restraint of reason, and luxuriate in the strangest conceptions. I have received letters about the comets of the last few years, enough to make one's hair stand on end at the absurdity of the theories they propose, and at the ignorance of the commonest laws of optics, of motion, of heat, and of general physics they betray in their writers. This is always the case whenever a great comet appears, only that in the later instances one feature of the general commotion of mind they inspire has been wanting. Thanks to the prevalence of juster notions of the constitution of the universe, and of the relation in which man stands to its Author; countries calling themselves civilized appear not to have been disgraced by any of those panic terrors, or thought it necessary to propitiate Heaven by any of those superstitious extravagances, about which we read on several former occasions. Even at Naples, which seems to be almost the lowest point of Europe in the scale of intellectual and social progress, I have not heard that it was thought necessary to liquefy the blood of St Januarius, or to carry his bones about the streets on account of any of these later great comets.

(2.) When we look through nature and observe the manifest indications of design which every point of it exhibits, it would be very presumptuous in us to assert that comets are of no use, and serve no purpose in our system. Hitherto, however, no one has been able to assign any single point in which we should be a bit better or worse off, materially speaking, if there were no such thing as a comet. Persons, even thinking persons, have busied themselves with conjectures: such as that they may serve for fuel for the sun (into which, however, they never fall), or that they may cause warm summers—which is a mere fancy—or that they may give rise to epidemics, or potato-blights, and so forth. But I need hardly say this is all wild talking, as my readers will be better able to judge when I shall have stated a few things which are known for certain about them. But *there is* a use, and a very important one, of a purely intellectual kind, which they have amply fulfilled; and who shall say that it has not been *designed* that such should be the case? They have afforded some of the sublimest and most satisfactory verifications of our astronomical theories—they have furnished us with a proof amounting to demonstration of the existence of a repulsive force* directed (under certain circumstances, and acting on certain forms of matter) *from* the sun as well as of that

* See on this subject my "Results of Astronomical Observations at the Cape of Good Hope," p. 407, *et seq.*, where the existence of such a repulsive force is clearly demonstrated.

great and general attractive force which keeps the planets in their orbits—and they have actually informed us of the *weight* of one of the planets which could not have been determined with any exactness if a comet had not on one occasion passed very near to it.

(3.) The ancients believed comets to be much of the same nature as meteors or shooting stars—either *in* the earth's atmosphere—not far above the clouds; or, at all events, much lower than the moon—or else as a species of vapours or exhalations raised up from the earth by the sun's heat, or by some other unknown cause; but they never for a moment dreamed of their forming part and parcel of that vast system of planetary bodies circulating about the sun, of which in fact they had hardly any distinct notions. In ancient history, however, several very remarkable comets stand recorded. One is mentioned by the Greek philosopher Aristotle in 371 B.C., with a tail extending over a third part of the sky. Many great comets are recorded at even more ancient dates in the Chinese annals: for that strange people kept an official record of all the remarkable stars, meteors, and other celestial appearances, for more than a thousand years before the Christian era, and what is stranger still, that record has been handed down to us and seems dependable. A great comet was seen close to the sun 62 years before Christ, during a total eclipse—and one which appeared in the year 43 B.C., soon after the murder of Julius Cæsar at Rome, was seen by all the assembled people in full daylight. Such a thing, though very uncommon, is by no means singular—it has happened several times, and in one case quite recently; for the great comet of 1843 was seen at noonday quite close to the sun both in Nova Scotia and at Madrid, and before sunset at the Cape of Good Hope.* Of course it is only the brightest part, or the head of a comet that can ever be so seen. The faint light of the tail has no chance of contending against broad daylight.

(4.) Before the invention of telescopes the appearance of a comet was a rare occurrence, because only a small proportion of them can ever be seen by the naked eye, and of them again only a small portion are considerable enough to attract much attention—but since that discovery it has been ascertained that they are very numerous—hardly a year passes without *one;* and very often two, three, and in one year, 1846, no less than eight were observed. Taking only two a year on an average as visible if looked for in a telescope, and considering that at least as many must occur in such situations that we could not expect to see them—in the 6000 years of recorded history there must have been between twenty and thirty thousand comets, great and small. A *great* comet, however, hardly occurs on an average oftener than once in fifteen or twenty years, or even yet more rarely;

* At Halifax, in the first mentioned colony, my informant saw a number of persons—natives of the place— hale and sturdy men, gathered in a group and gazing full on the sun, which, when he attempted to do, dazzled and almost blinded him. He was compelled to desist, and inquire what they were looking at, and how they could do so without being blinded. "Blinded!" was the reply—"Lord bless you, it does not hurt us;—what, can't you see it—that thing up by the sun?"

though, as sometimes happens in matters of pure accident and in the run of chances, it is not *very* unfrequent (and we have lately seen it remarkably exemplified) for two or even three very great comets to follow each other in rapid succession. Thus the great comet of 1680 was followed in 1682 by two other very conspicuous ones, of which we shall have more to say presently.

(5.) When a comet is first discovered in a telescope it is for the most part seen only as a small, faint, round, or oval patch of foggy, or, as it is called, *nebulous* light, somewhat brighter in the middle. By degrees it grows larger and brighter, and at the same time more oval, and at length begins to throw out a "*tail*," that is to say a streak of light extending always in a direction *from* the sun, or in the continuation of a line supposed to be drawn from the place of the sun below the horizon to the head of the comet above it. As time goes on, night after night the tail grows longer and brighter, the "*head*," or nebulous mass from which the tail seems to spring also increases, and within it begins to be seen what is called a "*nucleus*" or kernel, a sort of rounded, misty lump of light dying off rapidly into a haziness called the "*coma*" or *hair*. Within this, but often a good deal out of the centre, there is seen with a good telescope and a high magnifying power a very small spark or pellet of light which may or may not be the solid body of the comet, and which is the real nucleus. What in an indifferent telescope looks like a rather large puffy ball, more or less oval, is certainly not a solid substance. All the while the comet is getting every evening nearer and nearer to the place of the sun, and is therefore seen for a shorter time after sunset—or before sunrise, as the case may be—(for quite as many comets are seen in the morning before sunrise as in the evening after sunset). At last it approaches so near the sun as to rise or set very nearly at the same time, and so ceases to be seen except it should be so very bright and so great a comet as to be visible in presence of the sun.

(6.) When this has taken place, however, the comet is by no means to be considered as dead and buried. After a time it reappears, having passed by the sun, or perhaps before or behind it, and got so far away on the other side as to rise before the sun or set after him. If it first appeared after sunset in the west, it will now reappear in the east before sunrise. And what is very remarkable, its shape and size are usually totally different after its reappearance from what they were before its disappearance. Some, indeed, never reappear at all. The path they pursue carries them into situations where they could not be seen by the same spectators who saw them before. Others—like those which appeared in 1858 and 1861, without altogether disappearing as if swallowed up by the sun—after attaining a certain maximum or climax of splendour and size die away, and at the same time move southward, and are seen, as that of 1858 was (on the 11th of October for the first time), in the southern hemisphere, the faded remnants of a brighter and more glorious existence of which we here witnessed the grandest display. And on the other hand we here receive as it were many comets from the southern sky, whose greatest display the inhabitants of the southern parts of the earth only have witnessed. It also very often happens that a comet, which before its disappearance in the sun's rays was but a feeble and insignificant object, reappears magnified and glorified, throwing out an immense tail and exhibiting every symptom of violent excitement, as if set on fire by a near approach to the source of light and heat. Such was the case with the great comet of 1680—and that of 1843, both of which, as I shall presently take occasion to explain, really did approach extremely near to the body of the sun, and must have undergone a very violent heat. Other comets, furnished with beautiful and conspicuous tails before their immersion in the sun's rays, at their reappearance are seen stripped of that appendage, and altogether so very different that, but for a knowledge of their courses, it would be quite impossible to identify them as the same bodies. This was the case with the beautiful comet of 1835–6, one of the most remarkable comets in history. Some, on the other hand, which have escaped notice altogether in their approach to the sun burst upon us at once in the plenitude of their splendour, quite unexpectedly, as did that of the year 1861.

(7.) I come now to speak of the paths described by comets in the sky among the stars (which I need hardly observe keep always the same relative situations one among the other, and stand as landmarks, among which comets, planets, the moon and the sun pursue, or seem to us to pursue, their destined courses). Now we all know that the sun, moon, and planets, keep to certain high roads, like beaten tracks in the sky, from which they never deviate beyond definite and narrow limits assignable by calculation. With comets it is far otherwise. They are wild wanderers, and care nothing for beaten tracks. A comet is just as likely to appear in any one region of the starry heavens as in any other. They are no respecters of boundaries. The first time a comet is seen, no one can tell where it may next day be. The next observation still leaves a great uncertainty as to its future course. The third nails it. After three good observations, carefully made, of its place, we can thence foretell where it will go. Meanwhile, such is the variety of which their paths are susceptible, that for a very long time their movements were considered to be altogether capricious and unaccountable—creatures of chance—governed by no laws. Now the case is different. Most persons will remember that the comet of 1858 passed on the 5th of October of that year close to a very brilliant star, Arcturus, which shone through its tail at a very little distance from its root or outspring from the head. Well! within a very short time from the first appearance of that comet, while yet it was but a faint object, it was known to calculating persons that it *would* pass over Arcturus—the day—the hour—nay, almost the minute when the nucleus of the comet would be closest to the star were predicted—and the prediction was exactly verified. How this could happen I must now proceed to explain; but before I do so, I must premise that my hearers are not to be startled if I use some words that are not familiar to many of them, and ask for a little more of their attention than if I were merely telling some amusing story. What I am going to say will be already well known to a portion of them, but will be quite new to many,—and I will try to put it in such a way as shall not only be clearly intelligible, *but shall stick by them*, and become part and parcel of their minds and thoughts henceforward—and I am mistaken if many of this class of hearers (provided they will give me the attention the thing requires) do not rise from the perusal of this brief statement with much larger and higher conceptions of the magnificent system we belong to than they commenced it with.

(8.) The sun, as we all know, or may have heard, stands immovable, or nearly immovable, in the centre of our system, and all the planets, including the earth, circulate or revolve round it, each in its own time and at its own proper distance. These distances, for each planet, stand to each other in relations of proportional magnitude, which have become, by a long course of astronomical observation and calculations, known to us with extreme exactness, so that if the exact distance of any one of the planets from the sun, or the exact interval between any two of their orbits, can anyhow be ascertained in miles, yards, or feet, the dimensions of all the rest in similar units of measure may thence be derived. Supposing, for instance, we knew exactly the interval between the orbits of the earth and Mars, then if we would know the respective distances of the several planets in their order from the sun, it would only be necessary to multiply that interval, in the case of Mercury, by the decimal fraction 0·7392; in that of Venus by 1·3812; of the Earth by 1·9095; of Mars by 2·9095; of Jupiter by

9·9349; of Saturn by 18·2146; of Uranus by 36·6293; and of Neptune, the most distant of the known planets, by 57·3551.*

Now the interval between the earth's orbit and that of Mars (or the distance between that planet and the earth when they approach nearest) has quite recently been ascertained by a concerted system of observation, made during the past year, in which the astronomers in all the principal observatories of the globe have borne a part, and of which the final result has only within these few weeks become known. From these observations, so far as they have as yet been communicated and reduced,† it has been concluded that the interval in question is 6071 diameters of the earth, and as we know to a great nicety, by actual measurement of the earth's circumference, that its diameter is 7912½ miles, we are enabled at once to reduce the distance so obtained into miles (which gives 48,036,200 miles), and thence, as above indicated, to derive the earth's distance from the sun, which comes out 91,718,000, or about 92 millions of miles; and in the same way we may obtain the numerical dimensions in miles of the orbits of all the other planets, as also the sun's actual diameter, which appears to be 852,600 miles.

(9.) Such of our readers as may take the trouble to compare the distances and dimensions here set down with

* We consider in this and what follows, the orbits as circles, which is quite sufficient for purposes of illustration.

† Some time will probably elapse before our whole series can be collected and finally reduced.

those stated in my paper on the sun in the last lecture, will not fail to observe that they are materially smaller—by one thirtieth part of their respective amounts. The numbers there stated are in accordance with the state of our knowledge accepted at the time when that lecture was delivered, which rested for its basis on observations made upon Venus at the time of her transit across the sun's disc in the year 1769—observations by which the nearest distance of the orbits of Venus and the earth was concluded in terms of the earth's diameter, on the same general principle, though by a somewhat more refined and circuitous process, as that from which the least distance of Mars has just now been derived. As the circumstances of this earlier determination (delicacy of instruments and means of observation alone excepted) were much more favourable to exactness, astronomers would have hesitated in accepting the more recent conclusion in preference to the former, were it not for the support and corroboration it derives from another determination, also quite recent (though somewhat prior in point of date), depending on a direct measurement of the velocity of light by a peculiarly ingenious and delicate process invented and executed by M. Foucault. To explain the nature of this process here would lead me too far away from the immediate object of this discourse, from which, indeed, the whole of what is above said on the distance of the sun and planets would be justly considered as a digression were it not in some sort obligatory on every one to account for a departure from numerical statements once made. Suffice it therefore to say that the velocity of light so concluded was found to be somewhat less (and that by about one 30th part) of that which had been hitherto received (192,000 miles per second) and which was concluded from the observed *fact* of its traversing the diameter of the earth's orbit in $16^{m.}$ $26^{sec.}$ of time, and very considerably less than that before obtained by M. Fizeau, with a less perfect apparatus, and a less delicate and refined system of procedure. Now it will not fail to be remarked, that the *time* ($16^{m.}$ $26^{sec.}$) remaining unaltered, and the velocity *diminished* by one 30th, the distance traversed (the diameter of the orbit) in that time will also be diminished by the same aliquot fraction, so that there is a coincidence between the two corrections of the sun's distance, which, coming simultaneously, from such very different sources, cannot but lead to their acceptance, at least provisionally, and until the recurrence of that grand phænomenon, the transit of Venus, which will take place in the year 1874, shall put an end to all uncertainty on the subject of the true numerical dimensions of our system.

(10.) Bearing now these dimensions in mind, let us construct in imagination a figure consisting of concentric circles, to represent the orbits of the planets. Taking the largest, that of Neptune, as 30 feet in diameter, then will that of Uranus measure a little more than 19 feet across, of Saturn somewhere less than 10, of Jupiter rather more than 5, of Mars about 18 inches, and of the earth a foot, while the enormous body of the sun will stand represented in the centre of all by a pellet of very little more than one-ninth of an inch in diameter—the orbits of

Mercury and Venus by circles of 4½ and 9 inches respectively—that of the moon above the earth by one 15th of an inch, and the globe of the earth itself by a dot barely the thousandth part of an inch in size.

(11.) Strictly speaking, the orbits are not circles—they are slightly oval, or, as it is called, elliptic in form, and the sun does not occupy their common centre, but what is called the *focus* of each; that is to say, one of the two pins round which an ellipse may be described by carrying a pencil round them confined by a looped string encircling them both. The planetary orbits, moreover, all lie nearly in one plane, or very slightly inclined to that in which the earth performs its annual revolution, which is called the plane of the ecliptic—the angle at which the plane of each orbit meets and cuts this, being called its *inclination* to the ecliptic. They all circulate the same way round the sun, and the farther they are from the sun the slower they move—so that while the earth goes round it in 365 days, Mercury occupies only 88 in its revolution, while Neptune requires no less than 168 *years* to complete one of his circuits.

(12.) When we come to the comets, however, we find a very different state of things. A comet, it is true, moves round the sun as his centre of motion: not, however, in a circle, or any approach to a circle, but (with a very few, and those highly remarkable exceptions) in an immensely elongated, or, as it is termed, a very eccentric ellipse. In consequence, the nearest distances to which they approach the sun bear almost universally an exceedingly small proportion to those they attain when most remote, that is to say, at the two extremities of their elliptic orbits, or what are termed their perihelion and aphelion. By far the great majority approach it at their perihelion near enough to arrive within the earth's orbit —very many within that of Venus, or even of Mercury —and not a few attain an extreme proximity to the actual surface of the sun, while on the other hand only four or five among the vast number of recorded comets (those of 1747, 1826, 1835, 1847) have failed to arrive within twice the earth's distance, or within the orbits of those small planets called asteroids; and one only has had a perihelion distance exceeding four times the earth's distance (that of 1729), still falling short of the orbit of Jupiter. Probably, however, a comet, which should always remain outside of the latter planet's orbit, would have no chance of ever being seen by us. As to the extreme distances to which they recede from the sun, it is only in comparatively few instances that it can be even estimated—their ellipses being in general so elongated as to be undistinguishable from that extreme and limiting form which is called a parabola, which never returns into itself at all. The form of this curve is that which a stone thrown into the air describes, or which a jet of water thrown up obliquely by a smooth round pipe assumes in the air, being very much curved or bent about the point which is called the vertex, and less and less so in the ascending and descending branches.

(13.) Comets, we have said, are wild wanderers, and despise beaten tracks. No way confined, as the planets are, to move in planes nearly coincident with the ecliptic, they cut across it at every possible angle, and, as nearly as can be ascertained (with exception of one small class of comets), quite indifferently as to the degree of their *inclination*, or to the direction of the *longer axes* or longest dimensions of their orbits in space; so that there is no region of space, however situated either in direction or distance from the sun, which a comet may not visit. Neither do they conform to that other universal planetary rule of circulation round the sun in one direction. *Retrograde* comets, or those whose motion is opposite to that of the planets, are as common as *direct* ones, or those which conform to the planetary rule. Here again, however, there is a small class in which a tendency to conformity is exhibited, co-extensive with that above noticed, which affects a certain proximity to the ecliptic. But of this we shall have occasion to speak more at large.

(14.) It is only when all the particulars which determine geometrically the situation and the form of the orbit of a comet, its nearest distance from the sun, and the direction in which it is moving,—or what are called the *elements* of its orbit,—that it can be ascertained whether it has ever been seen before, and whether we are to expect ever to see it again; and that its future course, while it remains invisible, can be predicted with certainty. These elements are technically called—

1. The *perihelion* distance, or nearest approach to the sun.
2. The *eccentricity* of its ellipse, or whether the orbit be sensibly a parabola.

3. The *inclination* of its plane to the ecliptic.
4. The *longitude* of its node, or the direction of the line in which its plane intersects the ecliptic, which is called the *line of its nodes*.
5. The *longitude of its perihelion*, or, which comes to the same thing, the angle which the axis of the orbit makes with the line of nodes.
6. The exact moment when the comet passed through its *perihelion*, or was nearest to the sun.
7. The direction of its motion (direct or retrograde).

(15.) It is natural to ask *how* all these particulars ever *can* be known; and to this the answer is—By the same system of observation and calculation combined, by which we have come to know the form and dimensions of the orbits of the planets, their times of revolution round the sun, and their situation in space.

(16.) I believe it was Tycho Brahè, a celebrated Danish astronomer, who first rose to the conception that comets are beyond the moon, and not mere exhalations. The appearance of a great comet in 1577 set him thinking about it, and he was led by his observations and reasonings on them to a certain knowledge of the fact of its being much more remote than our own satellite; and he was therefore led to conjecture that the motions of comets had reference rather to the sun as their centre than the earth. The elliptic form of the planetary orbits was not then known, and Tycho accordingly supposed that comets moved about the sun in perfect circles. Borelli, a Neapolitan mathematician, suggested the idea of a long ellipse or parabola as the possible form of a comet's orbit; and Dörfel, a German astronomer in 1681, upon a careful consideration of all the observations of the great comet of 1680, came to the positive conclusion that that comet did really move in a parabolic orbit with the sun in its focus. This was an immense step; but neither Dörfel nor any one else could at that time give any account of the *reason why* this should be the case, or in what *manner* the comet was made to conform its sweep through space in so singular a way to the sun.

(17.) The wonderful discoveries of Sir Isaac Newton made all this clear. He first showed that the sun controls the movements of these wanderers by the very same force acting according to the very same law which retains the planets in their paths—that marvellous law of gravitation—the same power which draws a stone thrown from the hand back to the earth (in a parabolic curve)—which keeps the moon from flying off, and holds her to us as a companion—which keeps the planets in their circles, or rather ellipses, about the sun—and which we now know holds together several of the stars in couples, circulating one about the other.

(18.) The great comet of 1680, which occurred while Newton was brooding over these grand ideas which broke upon the world like the dawn of a new day in his "Principia," afforded him a beautiful occasion to test the truth of his gravitation theory by the most extreme case which could be proposed. The planets were tame and gentle things to deal with. A little tightening of the rein here and a little relaxation there, as they careered round and round, would suffice perhaps to keep them regular, and guide them in their graceful and smooth evolutions. But here we had a stranger from afar—from out beyond the extremest limits of our system—dashing in, scorning all their conventions, cutting across all their orbits, and rushing like some wild infuriated thing close up to the central sun, and steering short round it in a sharp and violent curve with a speed (for such it was) of 1,200,000 miles an hour at the turning point, and then going off as if curbed by the guidance of a firm and steady leading rein, held by a powerful hand, in a path exactly similar to that of its arrival, with perfect regularity and beautiful precision; in conformity to a rule which required not the smallest alteration in its wording to make it applicable to such a case. If anything could carry conviction to men's minds of the truth of a theory, it was this. And it did so. I believe that Newton's explanation of the motions of comets, *so exemplified*, was that which stamped his discoveries in the minds of men with the impress of reality beyond all other things.

(19.) This comet was perhaps the most magnificent ever seen. It appeared from November 1680 to March 1681. In its approach to the sun it was not very bright, but began to throw out a tail when about as far from the sun as the earth. It passed its perihelion on December 8—and when nearest was only *one-sixth* part of the sun's diameter from his surface—one fifty-fourth part of an inch on the conventional scale of our imaginary figure, and at that moment had the astonishing speed I have just mentioned. *Now observe one thing.* The distance from the sun's centre was about one 160th part of *our* distance from it. All the heat we enjoy on this earth comes from the sun. Imagine the heat we should have to endure if the sun were to approach us or we the sun to $\frac{1}{160}$th part of its present distance. It would not be merely as if 160 suns were shining on us all at once, but 160 times 160, according to a rule which is well known to all who are conversant with such matters. Now that is 25,600. Only imagine a glare 25,600 times fiercer than that of an equatorial sunshine at noonday with the sun vertical. And again, only conceive a light 25,600 times more glaring than the glare of such a noonday! In such a heat there is no solid substance we know of which would not run like water—boil—and be converted into smoke or vapour. No wonder it gave evidence of violent excitement—coming from the cold region outside the planetary system, torpid and icebound; already when arrived even in our temperate region it began to show signs of internal activity—the head had begun to develop and the tail to elongate till the comet was for a time lost sight of. No human eye beheld the wondrous spectacle it must have offered on the 8th December. Only four days afterwards, however, it was seen: and its tail, whose direction was reversed and which (observe) could not possibly be *the same tail* it had before—(for it is not to be conceived as a stick brandished round, or a flaming sword, but fresh matter continually streaming forth),—its tail I say had already lengthened to an extent of about

90 millions of miles, so that it *must* have been *shot out* with immense force in a direction *from* the sun, *a force far greater than that with which the sun acted on and controlled the head of the comet itself, which, as the reader will have observed, took from November* 10 *to December* 8, *or* 28 *days, to fall to the sun from the same distance, and that with all the velocity it had on November* 10 *to start with.*

(20.) All this is very mysterious. We shall never perhaps quite understand it, but the mystery will be at all events a little diminished when we shall have described some of the things which are seen to be going on in the heads of comets under the excitement of the sun's action, and when calming and quieting down afterwards. At present, however, we must get on with another part of our subject.

(21.) Only two years after this appeared another brilliant comet, and our countryman, Edmund Halley, following Newton's example and employing his system of calculation, computed its orbit, assuming (which simplifies the calculation very much) that orbit to be a parabola. He found its path to be very different from that of Newton's comet. Instead of nearly grazing the surface of the sun, its nearest approach to it was about 55 millions of miles, or about half-way between the orbits of Mercury and Venus. The plane of its motion, too, was much less inclined to that of the planets' orbit or the ecliptic—viz., about 17¾, and its motion was not direct, as Newton's was, but retrograde.

(22.) Halley was encouraged by the good agreement of

his calculations with the observed places of this comet to collect observations of former comets, and endeavour to make out their paths, or, as we now express it, to determine the elements of their orbits. With incredible labour he calculated the orbits of twenty-four remarkable comets, and among them he found two whose "elements" agreed in a remarkable manner with those of his first comet—both great comets, viz., one observed by Appian in 1531, and one by Kepler in 1607, and he noticed also this fact, this remarkable approximate coincidence—from 1531 to 1607 is 76 years, and from 1607 to 1682 75 years. This led him to suspect that all three were one and the same comet, returning periodically; and guided by this idea he was led to examine the records of history for comets of earlier date. Among them, three turned up—in the years 1305, 1380, 1456—and when all these years are arranged in a series, you see that the intervals are alternately 75 and 76 years. This confirmed him in his impression of its periodical return; and emboldened him to predict its return about the end of 1758 or beginning of 1759. You will observe that he allowed *more* than an average length of the period (77 years) for the fulfilment of his prediction. He had a reason for this. He ascertained that in coming back it would pass near the planet Jupiter, which is a large and massive planet, and Newton's discoveries had already taught him to contemplate the possibility of some disturbance of its motion from the attraction of such a body, and even enabled him to perceive that it would act to *retard* the return or prolong the period. Such disturbances do really exist, and have often very considerable effects on the return of comets. This very comet, in the table of its returns set down in the note below,* offers some striking examples. There occurs, for instance, 1378 A.D. and not 1380 set down for one of the epochs of its appearance, with 78 years interval between that and 1456. The fact is that Halley was mistaken in supposing either the comet of 1305 or that of 1380 to be the same with that in question. That comet really appeared in 1378, but *that fact Halley had no means of knowing*. It has very lately come to light on searching the Chinese annals. And the same annals have informed us of no less than six other still more ancient appearances of this selfsame comet, the earliest in the 11th year before our Saviour. And this, it must be allowed, greatly tends to increase our confidence in those venerable records of Chinese history. All this apparent irregularity is owing to the action mainly of Jupiter, which is a general disturber of comets, and gives a vast deal of trouble to calculators, as I shall soon explain; and Saturn is not without a finger in the pie.

(23.) This prediction of Halley's, as the time for its accomplishment drew near, created a great sensation—all the astronomers furbished up their telescopes, and all the mathematicians set to work to calculate. The mutual actions of the planets in that long interval had been well studied, and it was clearly ascertained that

* A.D. 451, July 3; 760, June 11; 1378, Nov. 8; 1456, June 8; 1531, Aug. 24; 1607, Oct. 26; 1682, Sept. 14; 1759, March 12; 1835, Nov. 15.

Halley was right in his conjecture about Jupiter, and that in fact the return of the comet would be delayed by the attraction of that planet 518 days, and by that of Saturn 100 more, and that it would make its next closest approach to the sun within a month one way or another of the 13th of April 1759.

(24.) All the astronomers of Europe were looking out for it, eager to seize it on its first coming within the range of human vision. They were all disappointed of their prize. It was carried off by a Saxon farmer of the name of Palitzch, an astronomer of Nature's own creating, who was always watching the heavens,—without telescopes, without knowledge,—simply from the profound interest their aspect inspired him with. He it was who first caught sight of it, on the 13th December 1758. It was taken up by others and regularly observed. It passed its perihelion on the 13th of March, *just within* the limit of possible uncertainty the mathematicians had allowed for their calculations.

(25.) This was certainly a very great and signal triumph. It was repeated, with every circumstance that could make it decisive or give it notoriety, in the year 1835, the epoch of the next appearance of "Halley's Comet." The calculation of the planetary perturbations (as the disturbances they cause in each other's motions are called) had then been brought to great perfection. The passage through the perihelion was predicted by M. Pontecoulant to take place on the 12th November, and by Rosenberger between the 11th and 16th. In point of fact, it happened on the 15th. And this time, too, the astronomers were not beaten by the farmers. Their telescopes were from day to day pointed right on the spot where it would be sure to appear—which was advertised all over the world in the almanacs; and it was caught at the earliest possible moment, and pursued till it faded away into a dim mist.

(26.) When lost to European astronomers (for, like those of 1858 and 1861, it ran southwards), Mr Maclear and myself received it in the southern hemisphere; and it was fortunate we did so; for, extraordinary as were the appearances it presented on its approach to the sun, they were if possible surpassed by those it exhibited afterwards; and the whole series of its phænomena has given us more insight into the *interior œconomy of a comet* and the forces developed in it by the sun's action, than anything before or since.

(27.) When first it was seen, it presented the usual aspect of a round misty spot, and by degrees threw out a tail, which was never very long or brilliant, and which to the naked eye or in a low-magnifying telescope appeared like a narrow, straight streak of light, terminating in a bright head; which in a telescope of small power appeared capped with a kind of crescent; but in one of great power exhibited the appearance of jets, as it were, of flame, or rather of luminous smoke, like a gas fanlight. These varied from day to day, as if wavering backwards and forwards, and as if they were thrown out of particular parts of the internal nucleus or kernel, which shifted round, or to and fro, by their recoil, *like a squib not held fast.* The bright smoke of these jets, however, never seemed to be able to get far out towards the sun, but always to be driven back and forced into the tail, as if by the action of a violent wind setting against them,—always *from* the sun,—so as to make it clear that this tail is neither more nor less than the accumulation of this sort of luminous vapour darted off *in the first instance* TOWARDS the sun, as if it were something raised up, and, as it were, exploded by the sun's heat, out of the kernel, and then immediately and forcibly turned back and repelled *from* the sun.

(28.) As this comet approached the sun, its tail, far from increasing, diminished; and between the middle of November and the 21st of January, strange to say, both head (that is *coma*) and tail were altogether destroyed, or at least rendered invisible. On the 21st of January the comet was actually seen like a small star without any tail or any haziness, and was only known *not* to be a star by being exactly in its calculated place, and by its not being there next night. After that its head seemed to form again round this star, and grew rapidly and visibly from night to night, putting on appearances which could not be clearly apprehended without elaborate figures. This growth of the comet was so very rapid, that in the interval of 17 days from the time I first saw it as a round body its real bulk had increased to 74 times the size it then had—and at the same rate it continued to swell out, not, however, preserving a round form, but growing longer in proportion to its breadth as if it intended to develop a new tail. But this it never did—the dilatation or swelling out continued, and at one time it had exactly the appearance of a ground glass lamp—the light always becoming fainter and fainter, till it at last seemed to pass away from view from mere faintness. All this while, however, there was a sort of smaller and much brighter *interior comet* visible, with a tail-like appendage, which seemed to be as it were a conducting channel by which the matter of the newly-forming head was gradually retreating back into the centre.

(29.) The discovery of the periodical return of Halley's comets forms an epoch in the history of their bodies. Since that time a great many more have been ascertained to return at regular intervals. I will mention some of the most remarkable cases of this kind.

(30.) In 1770 a comet appeared which proved rebellious to the then adopted system of calculation, which set out with assuming the orbit to be a parabola. It very soon appeared, by the calculations of M. Lexell, that the real orbit was an ellipse, and that not a very eccentric one. In fact, all the observations were perfectly consistent with an ellipse nearly coincident with the plane of the earth's orbit, of such dimensions as that the extreme excursion from the sun would carry it over a little beyond the orbit of Jupiter, and its nearest approach would bring it within that of Venus—the time of its revolution being 5½ years. Here was quite a new fact. All other comets then known had run out to limits far beyond our system—since even Halley's, with its period of 76 years at its greatest distance from the sun, passed very far beyond the orbit of Saturn, the most distant planet then known, and in fact beyond the two since discovered, Uranus and

Neptune. But here we seemed to have quite a sort of tame comet keeping within bounds, and within call. Of course its return was watched for with eagerness, but alas! it never made its appearance again. At its next return in 1776 this was well accounted for, as owing to the relative situations of the earth, sun, and comet, it could not have been visible; but at the next, in 1781, the earth was favourably situated, since $5\frac{1}{2}$ years would place the sun in the opposite part of its orbit; but 11 years in the same, and the calculators for a time were puzzled. The solution of the enigma was a very strange one. The poor comet had got bewildered. It had plunged headlong into the immediate sphere of Jupiter's attraction—had intruded, an uninvited guest, into his family circle—actually nearer to him than his fourth satellite, and into a situation where Jupiter's attraction for it was two hundred times that of the sun. Of course its course was for a time commanded entirely by this new centre of motion, and the comet was completely diverted from its former orbit.

(31.) So far all was clear enough. But people began to ask how, with so short a period, and being a tolerably large comet, it had never been seen before? Here again Lexell called Jupiter to the rescue. As he had taken away, so it turned out he had given. Jupiter, it will be borne in mind, comes round to the same point of his orbit in 11 years and 10 months; two of the comet's revolutions would occupy 11 years and 3 months, so that tracing back the comet two revolutions in its ellipse, and Jupiter rather less than one in his circle from the place of their final rencontre, which took place in 1779, it is clear they could not have been far asunder in 1767, 3 years before it became visible; and in fact, on executing the calculations necessary, it was clearly proved that before 1767 this unhappy comet had been revolving in a totally different orbit of much greater dimensions, and was actually siezed upon then and there by Jupiter, flung as it were inwards—and then after making two visits to the sun, again seized on, and thrown off into space, into an orbit of 20 years' period, where perhaps it may be quietly circulating to this day. Jupiter, in fact, is a regular stumbling-block in the way of comets.

(32.) This is a strange history—but it proved a very instructive one. The comet passed, as I have said, through the system of Jupiter's satellites. Now the motions of these bodies have been studied with a degree of care and precision quite remarkable by reason of their furnishing one of the means for ascertaining the longitudes of places. And if the comet had been a heavy massive body, its attraction must have produced some sensible disturbance in their motions. But no, not a trace of anything of the kind was detected. One and all of them pursued their courses with the very same precision and regularity as if nothing had happened. The conclusion is irresistible. *That* comet at least had no sensible weight or mass—it was a mere bunch of vapours.

(33.) Another very remarkable periodical comet is that of Encke, which makes its circuit about the sun in 1200 days, or about 3 years and 4 months, in the same direction as the planets. It is but a small one, being seldom visible without a telescope. Its orbit was first computed on its appearance in 1795 (when it was discovered by Miss C. Herschel), and again in 1805 and 1819. Upon this last occasion M. Encke, an eminent computist, found that its motion could not be explained without supposing it to move in an ellipse of the last period I have mentioned—and on searching back into the records of comets he found those two I have just named, which agreed perfectly, and proved to have been really the same.

(34.) Since that time it has been re-observed on every subsequent revolution in '22, '25, '29, '32, '35, '38, '42, '45, '48, '51, '55, and is always announced in the almanacs as a regular member of our system. Its nearest approach to the sun brings it just within the orbit of Mercury, and on one occasion that planet happened to be so very near it on its arrival, that it produced a pretty considerable disturbance of the comet. But here, too, as in the case of Lexell's comet, not the smallest perceptible effect was produced *by* the comet *on* the planet; and thus two valuable pieces of information were gained. *First;* Astronomers were enabled to estimate the mass or weight of that small planet better than by any other means; and *secondly;* It was proved that *this* comet also has no perceptible weight—and is also a mere puff of vapour, or something as unsubstantial.

(35.) There is another strange fact which this comet has revealed. Its successive revolutions are each a little shorter than the last—a small fraction of a day, it is true, but still unquestionably made out. This has been held to prove that the comet is by very slow degrees approaching the sun, and will at last fall into it—as if it moved in a space not quite empty, and were in some very slight degree resisted in its motion. I cannot quite reconcile myself to this opinion, and I think I have perceived another explanation of the fact, which I have given elsewhere; but to state this would lead me too far, and I must now go on to relate one of the strangest and most uncouth facts of this strange cometic history.

(36.) On the 27th February 1826, Professor Biela, an Austrian astronomer of Josephstadt, discovered a small comet. When its motions were carefully studied it was found by M. Clausen, another of those indefatigable German computists, that it revolved in an elliptic orbit in a period of 6 years and 8 months. On looking back into the list of comets, it proved to be identical with comets that had been observed in 1772, 1805, and perhaps in 1818. Its return was accordingly predicted, and the prediction verified with the most striking exactness. And this went on regularly till its appearance (also predicted) in 1846. In that year it was observed as usual, and all seemed to be going on quietly and comfortably, when behold! suddenly on the 13th of January it split into two distinct comets! each with a head and coma and a little nucleus of its own. There is some little contradiction about the exact date. Lieutenant Maury, of the United States Observatory of Washington, *reported officially on the 15th having seen it double on the 13th*, but Professor Wichmann, who *saw it double on the*

15*th*, avers that he *had a good view of it on the* 14*th*, and remarked nothing particular in its appearance. Be that as it may, the comet from a single became a double one. What domestic troubles caused the secession it is impossible to conjecture, but the two receded farther and farther from each other up to a certain moderate distance, with some degree of mutual communication and a very odd interchange of light—one day one head being brighter and another the other—till they seem to have agreed finally to part company. The oddest part of the story, however, is yet to come. The year 1852 brought round the time for their reappearance, and behold! there they both were, at about the same distance from each other, and both visible in one telescope.

(37.) The orbit of this comet very nearly indeed intersects that of the earth on the place which the earth occupies on the 30th of November. If ever the earth is to be swallowed up by a comet, or to swallow up one, it will be on or about that day of the year. In the year 1832 we missed it by a month. The head of the comet enveloped that point of our orbit, but this happened on the 29th of October, so that we escaped that time. Had a meeting taken place, from what we know of comets, it is most probable that no harm would have happened, and that nobody would have known anything about it.*

* It would appear that we are happily relieved from the dread of such a collision. It is now (Feb. 1866) over due! Its orbit has been recomputed and an ephemeris calculated. Astronomers have been eagerly looking out for its reappearance for the last two months, when, according to all former experience, it ought to have

(38.) The number of comets whose periodical return has been calculated is pretty considerable. Altogether about 36; and of these there are 5 which revolve in periods of from 70 to 80 years, and several of the rest in short periods from 3 to 7 years; and it is a very remarkable feature in their history that *all* the comets of short period, and three out of the five of those of the larger ones specified, revolve in the *same* direction round the sun as the planets, and have their orbits inclined at no very large angles to the ecliptic.

(39.) Of comets not periodical, I have already mentioned that most remarkable one of 1680, but several others deserve special notice. That of 1744 was a truly wonderful object. It is described, and has been depicted, with six tails spread out like an immense fan—extending 30° from the head—which is fully the extent of the tail of the comet of 1858; and the appearance of its head when viewed through a telescope exhibited the same sort of jets of luminous smoke, the same curved envelopes and arches as I have already described, showing the same kind of excitement by the sun's heat, and the same action driving the vapour back into the tail.

(40.) The comet of 1843 was still more remarkable. Many of my hearers, I dare say, remember its immense

been conspicuously visible—but without success! giving rise to the strangest theories. At all events it seems to have fairly disappeared, and that without any such excuse as in the case of Lexell's, the preponderant attraction of some great planet. Can it have come into contact or exceedingly close approach to some asteroid as yet undiscovered; or, peradventure, plunged into and got bewildered among the ring of meteorolites, which astronomers more than suspect?

tail, which stretched half-way across the sky after sunset in March of that year. But its head, as we here saw it, was not worthy such a tail. Farther south, however, it was seen in great splendour. I possess a picture by Mr Piazzi Smythe, Astronomer-Royal of Edinburgh, of its appearance at the Cape of Good Hope, which represents it with an immensely long, brilliant, but very slender and *forked* tail. Of all the comets on record, that approached nearest the sun—indeed, it was at first supposed that it had actually grazed the sun's surface, but it proved to have just missed by an interval of not more than 80,000 miles—about a third of the distance of the moon from the earth, which (in such a matter) is a very close shave indeed to get clear off. There seems very considerable reason to believe that this comet has figured as a great comet on many occasions in history, and especially in the year 1668, when just such a comet, with the same remarkable peculiarity, of a comparatively feeble head and an immense train, was seen at the same season of the year, and in the very same situation among the stars. Thirty-five years has been assigned with considerable probability as its period of return, but it cannot be regarded as quite certain. (It will of course be understood that the return of a great comet to the neighbourhood of the sun by no means implies that it should be a conspicuous one, as seen from the earth. The phase of its greatest development may be, and is, indeed, more likely than not to be, ill-timed, as regards the relative situations of the earth and sun, for its exhibition as a great celestial phænomenon.)

(41.) Another great comet which has assumed a sort of historical and political importance is that which appeared in A.D. 1556. According to the account of Gemma, it would not seem to have been a very large one, as he assigns to it a tail of only four degrees long. Its head, however, equalled Jupiter in brightness, and in size was estimated at about one-third or one-half of the diameter of the moon. It appeared about the end of February, and on the 16th of March is described by Ripamonte as a really terrific object. Terrific indeed it might well have been to the mind of a prince prepared by the most abject superstition to receive its appearance as a warning of approaching death, and as specially sent, whether in anger or in mercy, to detach his thoughts from earthly things, and fix them on his eternal interests. Such was its effect on the Emperor Charles V., whose abdication of the imperial throne is distinctly ascribed by many historians to this cause, and whose words on the occasion of his first beholding it have even been recorded—

"*His ergo indiciis me mea fata vocant!*"

the language and the metrical form of which exclamation afford no ground for disputing its authenticity, when the habits and education of those times are fairly considered. This comet has been supposed to be periodical, and to return in 291 years, on the ground of the prior appearance of great comets in the years 975 and 1264 (at intervals, that is, of 289 and 292 years respectively), and the general agreement of their orbits, so far as could be made out from the imperfect records we possess of their courses, with that of the comet in question. The next return, on this supposition, would have fallen about the year 1846 or 1847. It did not, however, appear at that epoch, nor in any subsequent year up to the present time, although, from some very elaborate calculations by Mr Hind and Professor Bomme (too elaborate, it would appear, to have been bestowed on the imperfect records we possess of its previous history) it should have been delayed by planetary perturbations for several years beyond that date, and even so late as to the year 1858 or 1860.

(42.) Accordingly, when the three great comets, whose arrival in and since the year 1858 has so surprised and delighted the astronomical world, made their successive appearances, there were few persons at all acquainted with cometary history whose first impression was not that of the return of "Hind's Comet," as it had grown to be called, from the eminent calculator and mathematician who had bestowed so much pains on it. This, however, it is needless to observe, was not the case. Neither of them had ever been seen before, nor can either of them ever be expected to appear again, unless to a posterity which may look back on our record of them as we do on those ancient Chinese annals already spoken of. Of these, by far the most magnificent in point of mere display, as well as the most interesting, when contemplated in a physical point of view, was that of 1858 (the fifth of that year), or Donati's comet, as it is now called, from the astronomer of that name, who first observed it at Florence on the 2d of June, at which time it appeared only as a round misty patch or "nebula." This was about a month after it had passed from the southern to the northern side of the plane of the earth's orbit: and that of the comet being very highly inclined (63°) to the ecliptic; its perihelion lying also on the north side of that plane; its motion being retrograde, and the earth accordingly advancing to meet it;—all these favourable circumstances concurring, it so happened that our nearest proximity to it occurred only six days after its "perihelion passage" or time of nearest approach to the sun, which took place on the 29th of September, and in a situation with respect to the sun every way advantageous to obtaining a good view of it. Accordingly, with the exception of the comet of Halley in 1835, no comet on record has been watched with such assiduity, or been more thoroughly scrutinized. A *resumé* of all the observations of it has been recently published by Professor Bond, forming the third volume of the "Annals of the Observatory of Harvard College, in the United States," in which its appearance in every stage of its progress is represented in a series of engravings, which in point of exquisite finish and beauty of delineation leave far behind everything hitherto done in that department of astronomy.

(43.) It was not till the 14th of August, or 73 days after its first discovery, that it began to throw out a tail, and to become a conspicuous object. Very soon after this, its first appearance; a slight but perceptible curvature was perceived in the tail, which, on the 16th of September, had become unmistakable, and continued to increase in amount as the latter extended in apparent dimension, till it assumed at length that superb aigrette-like form, like a tall plume wafted by the breeze, which has never probably formed so conspicuous a feature in any previous comet. To a certain extent, it is a common enough feature in the tails of comets, and is usually regarded as conveying the idea of their moving in a resisting medium;—in a space, that is to say, not quite empty, as smoke is left behind a moving torch. But this is a very gross and inadequate conception of the peculiarity in question. The resistance of the "ether," such as the phænomena of Encke's comet already noticed, may be supposed to indicate, is far too infinitesmally small to be competent to produce any perceptible deviation from straightness. Nor is it at all necessary to resort to any such explanation of the fact. Such an appearance would naturally arise from a combination of the motion the matter of the tail had (in participation with that of the nucleus) with the impulse given it by the sun—each particle of it describing, from the moment of quitting the head, an orbit quite different from that of the latter; being necessarily, under the influence of *the repulsive* force directed *from* the sun, a curve of the form called by geometers an hyperbola, nearly approaching to a straight line, and having its *convexity* turned towards the sun: the visible form of the tail (be it observed) being, *not the perspective view of such an orbit*, but that of the *portion of space* containing, for the time being, all those particles, each describing its own independent orbit, and each reflecting to the eye its quota of the solar light.*

(44.) A very striking feature in Professor Bond's engravings, which he describes as frequently and certainly observed in America, and which did not pass wholly unnoticed in Europe, consists in the appearance of one, and on some nights two, excessively faint, narrow, and perfectly straight rays of light, or "secondary tails," starting off from the main tail on its preceding or anterior side (that towards which the comet was advancing, and which side was always the brightest, sharpest, and best defined) in the direction of tangents to its curvature at points very near the head, and extending on some nights (on the 4th, 5th, and 6th of October) to a much greater length than the primary or more luminous tail. These appearances were presented from the 28th September to the 11th of October with more or less distinctness. They are peculiarly instructive, as they clearly indicate an *analysis of the cometic matter by the sun's repulsive action*—the matter of the secondary tails being evidently darted off with incomparably greater velocity (indicating an incomparably greater intensity of repulsive energy) than that which went to form the primary one. The primary tail also presented another feature, frequently, indeed almost always, observed in comets, viz.,

* Some anomalous appearances in the early development of the tail in this comet, which was slightly curved, even when the earth was in the plane of the orbit, can by no means be regarded as fatal to this explanation of the general phænomenon, as they might have originated in a lateral direction of projection of the caudal matter from the nucleus *in ipso motus initio.*

its separation, behind the head, into two main streams with comparative darkness between them. This would be a natural and necessary optical consequence of the tail consisting of a hollow, conical envelope, streaming off on all sides around from the head, and presenting to the eye therefore a much greater thickness of luminous matter at its edges than at its middle. But in this comet the separation, when viewed through powerful telescopes, was singularly sharp; and appeared as a clear, narrow, straight cut, or dark chink, originating close to the nucleus (as, indeed, on that explanation of the fact it ought). And this brings me to treat of the appearances presented by the head and nucleus under the inspection of powerful telescopes.

(45.) All considerable comets which have been examined with anything like what would in these days be regarded as a *powerful* telescope, have presented the appearance of a nucleus of more or less definable and condensed light, sometimes having a much brighter and almost stellar point in or near its centre, and at some distance, *in the direction of the sun*, a capping of light sometimes quite separated, as if some transparent atmosphere sustained it—more frequently connected by those fan-like jets of "flame," such as we have mentioned in the case of Halley's comet, and putting on the aspect of a "sector," or fan, opening out into a widening arc, and bounded internally by two crescents springing from the nucleus. Donati's comet exhibited this feature in perfection; not, however, without striking variations and individual peculiarities. There was the same appearance with *low magnifying powers* of an envelope surrounding a nucleus in the general way above described, but the connexion was singularly varied, as if several jets of luminous (or illuminated) matter had been issuing from various parts of the nucleus, giving rise, by their more or less oblique presentation to the eye, to exceedingly varied appearances—sometimes like the spokes of a wheel or the radial sticks of a fan, sometimes blotted by patches of irregular light, and sometimes interrupted by equally irregular blots of darkness. From the 24th September to the 10th October, however, there were seen to form no less than *three distinct* caps or envelopes in front of the nucleus, each separated from that below it by a more or less distinct comparatively dark interval. These Professor Bond appears to consider as having been thrown off in intermittent succession, as if the forces of ejection had been temporarily exhausted, and again and again resumed a phase of activity; the peculiar action by which the matter of the envelopes was ultimately driven into the tail (or, as we conceive it, an analysis of that matter performed by solar action, the *levitating* portion of it being hurried off—the *gravitating* remaining behind in the form of a transparent, gaseous, non-reflective medium), taking place, not on the surface of the nucleus, but at successively higher levels. Meanwhile, and especially from the 7th to the 10th of October, that is to say, when the full effect of the perihelion action had been endured, the nucleus and its adjacent sector offered every appearance of most violent, and, so to speak, angry excitement, evidenced by the complicated structure and convolutions of the jets issuing from it. From this time, to its final disappearance, the violence of action gradually calmed down, while the comet itself went southwards, and at length vanished from our horizon.

(46.) An idea of the actual dimensions of this comet may be formed from the measurements taken by Professor Bond on the 2d October, which, combined with the distance of the comets from the earth at that date afford the following results, viz.:—

	Miles.
Diameter of the bright internal pellet or nucleus, .	1,600
Distance from its centre to the summit of the first envelope, ...	7,500
Distance to that of the second envelope, . .	13,200
Breadth of the brightest part of the tail where it seemed (to the naked eye) to issue from the comet, ...	90,000

to which it may be added that the actual length of the tail, when at its greatest development, could not have been less than 30 millions of miles, and those of the faint streaks or secondary tails 34 or 35 millions.

(47.) The comet of 1861, which burst suddenly on us in its full splendour on the 30th of June in that year (though it had been seen for seven weeks before in the southern hemisphere), was considered by those who saw it at its first appearance to surpass in brightness even that of 1858, and was remarkable for the extreme breadth and diffusion of its tail when first seen, arising from the circumstance of the earth having been then situated nearly in its prolongation. Indeed, it is not impossible that on that day we actually traversed some portion of it, our distance from the head being then only about 13,000,000 miles, and more than one observer having noticed and been much struck with an unusual and general brightness, like an auroral light not confined to the neighbourhood of the comet, but spreading over the whole sky. The most remarkable peculiarity of this comet, however, consisted in the enormous length which one side of its tail attained between the 2d and the 4th of July, extending in a *perfectly straight* but feeble ray from near the star Alpha in the Great Bear, to and beyond that designated by the same letter in Ophiuchus, or over 75 degrees in angular measure, contrasting strikingly with the stunted development and *bushy* aspect of the opposite branch. Its head, when viewed with good telescopes, exhibited the same general phænomena of luminous jets and crescent-like emanations as its predecessor, but much less complex and varied. Owing to the great inclination of its orbit, this comet, coming to us from the southern side of the ecliptic, soared high above it on the northern side and remained long and conspicuously visible as a *cricumpolar* object, the whole of its diurnal course being above our horizon. Not so its successor of 1862, whose orbit being but slightly inclined to our own, its motion retrograde (or meeting the earth), its perihelion distance almost exactly equal to our distance from the sun, and its passage through the perihelion occurring at a time when the earth was not very remote from that point, it passed us closely and swiftly, swelling into importance, and dying away with unusual rapidity. The phænomena exhibited by its nucleus and head were on this account peculiarly interesting and instructive, it being only on very rare occasions that a comet can be closely inspected at the very crisis of its fate, so as to witness the actual effect of the sun's rays on it. In this instance, the pouring forth of the cometic matter from the singularly bright and highly condensed, almost planetary nucleus, took place in a single compact stream, which after attaining a short distance, equal to rather less than a diameter of the nucleus itself, was so suddenly broken up and dispersed as to give, on the first inspection, the impression of a double nucleus. The direction of this jet varied considerably from day to day, but always declined more or less in one direction from the exact direction from the sun. So far as I am aware, the formation of an envelope disjoined from the head was not witnessed in this comet.

(48.) And now, I daresay, all my hearers are ready to ask—After all what *is* the tail of a comet? Is it material substance in the first place? To this I answer unhesitatingly, Yes! Donati's comet has given a decisive proof on that point. There is a criterion by which, *when it is observed,* it can be positively asserted that the light by which anything is seen has been reflected from a material substance. The light reflected, when it exhibits that peculiar property in which this criterion consists is said to be *polarized.* The direct light of the sun or that of a candle is not polarized, but when reflected at a particular angle on any surface but a metallic one, it *is*, and *if* it is polarized, we may be sure that it is not direct light thrown out by the object seen, but borrowed or indirect light. No matter at present what this polarization is, all I wish to convey is, that there is a simple enough experiment which everybody who understands optics knows how to make, which *if the result be of a certain kind*, the reflection of the light is demonstrated—(the converse, be it observed, does not hold good)—in an instant, by merely looking through a small instrument contrived on purpose. Now, Mr Airy, the present astronomer-royal, a person who is not only an excellent astronomer, but who stands very high as an authority on this especial branch of optics, applied this test to the light of the comet's tail on the 27th September, and found it polarized. The tail then shone by reflected light, and there was also another particular indication or character of the polarization impressed, which the same trial afforded, and which enabled him to say positively that the light had been reflected from some source of light agreeing in situation with the sun.

(49.) The tail of the comet then was material substance.* But now, only conceive what must be the thinness, the almost spiritual lightness of a vapour or fog, which, occupying such an enormous space, would not extinguish

* I applied the same test to the comet of 1862. There are various modes of making the trial. Mine was by looking at the comet through an achromatized doubly refracting prism, and turning the prism round in its own plane. I could perceive *no* alternate maxima and minima of brightness in the images. But in this case it is the positive result which is conclusive. Everything depends in the first instance on the relative situations of the objects and the eye. And, moreover, the light of the comet of 1862 was far inferior to that of Donati's, rendering the experiment *pro tanto* more delicate—and it is very possible that to septuagenarian eyes, indications of *partial polarization* might escape observation.

the stars shining through it. Arcturus was noway dimmed when it shone through the very middle of the brightest part of the tail of that comet. But I have already stated that that part measured 90,000 miles, and as this part of the tail was no doubt *round*, as thick as broad, the star's light must have shone through 90,000 miles of this mist. Now, every one must have noticed that the steam puff of a railway carriage completely obscures the sun, much more a star. You cannot see the sun through it. Well, then: there must have been less substance in the line of 90,000 miles of tail between the eye and star than in the line of a few yards of steam smoke penetrated by the eye in the other case.

(50.) If you look at a filmy cloud at sunset, though not thick enough to hide a star, you see it bright with vivid golden light by reflection from the sun. How much more then if it were much nearer to the sun, and much more strongly illuminated. Such a cloud is penetrated with light through its whole thickness and reflects it equally from its interior and exterior. Just so in the almost infinitely more thin texture of a comet—even in the densest part of the head it cannot be compared to the lightest cloud so far as substance goes. In Biela's comet very minute stars have been seen by myself through a part of the head at least 50,000 miles in thickness, which a fog a few yards thick would have extinguished. A solid body of a round shape would exhibit phases like the moon, and would appear sometimes as a half moon, sometimes as a crescent, and sometimes as a full moon—but the heads of comets show no such appearances. Of course I do not mean to deny that that very minute brilliant point which some are said to have exhibited, may not be a solid body—but it must be a very small one—perhaps not a tenth or a hundredth part the size of the moon; and, indeed, if there be *not* some little solid mass, it seems impossible to conceive how the observations of a loose bundle of smoke, rolling and careering about, could ever be represented by any calculation. Certain it is, that what appears to be the central point of a comet, *is* that point (and no other is) which conforms rigorously to the laws of solar gravitation, and moves strictly in a parabolic or elliptic orbit.

(51.) There is a very curious feature common to all the comets which have little or no tail, and which circulate about the sun in short periods; such as that of Encke, in which it has been especially observed. As they approach the sun, so far from dilating in size, they contract,—I mean in their real bulk, orat least their visible bulk,—and on receding from the sun they grow again to their former size. The only possible explanation of this is, that a portion of their substance is evaporated by the heat—that is to say, converted from the state of fog or cloud into that of invisible transparent vapour. Perhaps I ought to explain what is the difference. Take the case of a light cloud in a clear sky when the sun shines on it. If you watch it attentively, you will very often see it grow thinner and thinner, and at last disappear altogether. It has been converted from *mist* to invisible *vapour.* The material substance, the watery particles are *there,* but they have passed into another form of existence, in which, like the air itself, they are invisible. As the comet then gets heated a portion is actually vaporized—and the vapour condenses as it cools again. The whole substance of the comet of Halley, as you have heard, was so evaporated in 1835–6, all but what I suppose must have been really its solid body; that *star* which I have already mentioned, which was seen on the 22d January 1836: and all that curious process that went on afterwards, no doubt was that of the re-condensation of the evaporated matter, and its gradual re-absorption into and close around the body.

(52.) There is still one point in the history of comets which I have not touched upon, or but slightly. Comparatively only a few of the great number of comets which have been observed, and of which the orbits have been calculated, have been seen more than once—the great majority once seen, seem lost for ever. What becomes of them, is a very natural question. The answer to this is, that the time of the periodical return of a comet depends entirely on the distance to which it may run out from the sun. Now we know of nothing to interfere with or disturb the motion of a comet, once clear of the planetary system, between the farthest planet and the nearest fixed star; and that interval is so immense that the imagination is lost in attempting to conceive it. The farthest planet we know of is only 30 times the distance of the earth from the sun. Halley's comet in its elliptic orbit of 75 years, goes only a little beyond that, or to about 36 times the earth's distance. Donati's comet, if the computists are right, will return in 2100 years, and will have gone out to a distance 238 times the earth's distance from the sun, or nearly 80 times the distance of the planet Neptune. But this is still hardly the thousandth part of the distance to the very nearest fixed star —and supposing the elliptical orbit of a comet should be so long as to carry it out only half-way to the nearest star—its return to the sun would require upwards of 11 millions of years from its last appearance. Few of those who saw the last-mentioned comet pass over Arcturus, had any idea of the enormous distance at which the star really was behind the comet: and Arcturus is by no means the nearest star.

(53.) I think, from what I have said, you will perceive that there is in the history of comets matter enough both to encourage inquiry and to check presumption. Looking to the amount of our positive knowledge of them—knowledge acquired by centuries of observation, and by the conspiring efforts within the last two centuries of the profoundest thought and the most persevering labour of which man is capable, we may reasonably enough congratulate ourselves on what has been done, and while we can afford to look back with an indulgent smile on the unfledged and somewhat puerile attempts of the ancient mind to penetrate their secret, we may as reasonably look forward to the revelations they will afford, as time rolls on, of facts and laws of which at present we have no idea. This may, and ought to inspire confidence of the powers of man to penetrate always deeper and deeper into the secrets of nature. But, on the other hand, here, as on every other occasion, we find that the last and greatest discoveries only land us on the confines of a wider and more wonderfully diversified view of the universe; and have now, as we always shall have, to acknowledge ourselves baffled and bowed down by the infinite which surrounds us on every side.

(54.) Beyond all doubt, the widest and most interesting prospect of future discovery which their study holds out to us, is that distinction between *gravitating* and *levitating* matter, that positive and unrefutable demonstration of the existence in nature of a repulsive force, co-extensive with but enormously more powerful than the attractive force we call gravity, which the phænomena of their tails afford. This force cannot possibly be of the nature of electric or magnetic forces.* These forces are especially polar in their action between particle and particle—a magnet, or an electrified particle, of indefinitely minute dimensions—so minute as the discrete particles which go to form a comet's tail, could by no possibility be either attracted or repelled, *as such*, by a body, however powerfully magnetized or electrified, placed at the distance of the sun. It might have a *direction* given to its magnetic or electric *axis*, but its centre of gravity would not be

* This and much of what follows may seem inconsistent with what is said in my "Results of Ast. Obs., &c., at the Cape of Good Hope," p. 409, and note thereon. To a certain extent it is so, and to that extent it is a recommendation, but I am here speaking only of that portion of the matter of the comet whose chemical union may be considered as completely overcome, and whose levitating or negative constituent is fairly driven off, never to return. That which may be conceived to remain behind may conform under the circumstances of the case to the dynamical relations there indicated.

affected one way or the other. The attraction on one of its sides would precisely equal the repulsion on the other. The separation of one portion of the matter of a comet from the other by the action of the sun, which we see, unmistakably, operated at and near the perihelion passage (a separation which the late Sir William Herschel certainly *had* in mind, though perhaps somewhat indistinctly, when he spoke of a comet visiting our system for the first time as consisting of "unperihelioned" matter in contradistinction to those which he considered to have lost their tails by the effect of repeated appulses, and to consist mainly of *perihelioned* matter) —this separation I can only conceive, as I have ventured to express it above, as an *analysis* of the materials: analogous to that analysis or rather disunion by the action of heat which St Clair Deville has lately shown to take place between the constituents of water at high temperatures. In this latter case the chemical affinity is so weakened that the mere difference of difficulty in traversing an earthenware tube suffices to set them free of one another. How much more so, then, were the one constituent of a chemical compound subject to a powerful repulsion from a centre which should attract the other, and with it by far the larger mass of the total comet. Might not, under such circumstances, the mere ordinary action of the sun's heat sufficiently weaken their bond of union: and might not the residual mass, losing at every return to the perihelion more and more of its levitating constituents, at length settle down into a quiet, sober, unexcitable denizen of our system?

LECTURE IV.

THE WEATHER, AND WEATHER PROPHETS.

"Varium et mutabile semper."

"THERE is an ugly look about the sky, and the wind is getting up, and Fitzroy's storm-signals were hoisted yesterday evening and are up now. We shall have a gale. I am afraid we must put off our boating for to-day anyhow," said my friend A—— to his wife the other day; "there may be nothing in it; but we should look very silly to come home half-drowned in the face of a warning."

(2.) And it was well the lady took the advice. It was but a pleasure party after all. But the fishermen to whom the loss of a day was a serious matter, put off. Not that they altogether pooh-pooh'd the inverted cone and drum: but they reckoned on twenty-fours' law at least, and suffered for their miscalculation. One boat came on shore in fragments, several suffered damage, and all agreed it would have been wiser to have stayed at home.

(3.) An occurrence like this took place at one of our southern watering-places not far from hence, a few days ago; and the gale which followed was one of the precursors of that far more fearful one which has just (apparently) blown itself out;* part and parcel, no doubt, of that great periodical phænomenon whose recurrence under the name of "the November atmospheric wave," is beginning to be recognized as one of the features of our European weather table—a vast and considerably well-defined atmospherical disturbance; peculiar, it would seem, to this portion of the globe, though originating, as we shall see reason to believe, in the opposite hemisphere; and of which the gale of the Royal Charter (October 25, 1859); the great Crimean hurricane of disastrous memory (November 14, 1855); and the still more awful storm of December 8, (N.S.) 1703, the greatest which has ever swept this island,—may be considered as shadowing out the beginning, middle, and end.

(4.) The actual barometric fluctuation to which the epithet has been affixed by Mr Birt, who first drew attention to one of its most peculiar features, is, however, confined to narrower limits of time; and refers to one great billow or mountainous breaker (so to speak) of air, which sweeps in November across the whole North Atlantic and the European continent from N.W. to S.E.; preceded and followed by sudden and violent subor-

* This was written on the morning of the 2d of November 1863, after a night of most terrific storm.

dinate fluctuations, embracing in their whole extent and in different years the longer period referred to.*

(5.) Meteorology, so far as prediction of the weather is concerned (which most persons consider, very erroneously, to be its only practical object), may be regarded as a science still in its infancy; though if such be the case, to judge from the voluminous nature of its records, and the multitude of books which have been written on it, its maturity, if ever attained, would promise to be gigantic indeed; were it not that the progress of all real science is towards compression and condensation, and its whole aim to supersede the endless detail of individual cases by the announcement of easily remembered and readily applicable laws. Most of the indications of the "weatherwise," from Aratus down to Foster, have hitherto been little more than what, in the language of Mr Mill, would be called "simple connotations." The condor is circling in the sky: *therefore* a lion is devouring a horse below. The sheep turn their tails to the south-west: *therefore* there will be a gale of wind from that quarter. The "Rainbow in the morning," &c. The "Evening red and the morning gray," &c., &c. All such connotations have their value in an absolute ignorance of causes and modes of action: but it is only by the study of *these* that we learn what to connote. And there is no doubt, that since, after an immense

* This is the direction of the progress of the *wave*. That of the *wind* during the gales which accompany it is at right angles to that direction, or from S.W. to N.E. : in analogy (?) to the *transverse rotation* of the etherial molecules in the propagation of a circularly polarized ray of light.

amount of persevering labour bestowed on daily and hourly records of the weather, an insight (and no inconsiderable one) *has* been gained into the *causes which determine it*, and the sequence of phænomena which exhibit them in action; a style of connotation *has* commenced, which is already bearing practical fruit, in the form of telegraphic warnings of approaching bad weather, of positive value and interest. There can be no better proof of this, than in the fact that the example set by our own Admiralty in the establishment of a system of coast weather signals, has already been followed to a certain extent in Holland, and is in course of being so in France. Nations are perhaps not overready in following up the improvements of their neighbours; but at all events, they are remarkably slow in adopting each other's practical blunders.

(6.) The indications of the coming weather which experience has shown to be in any degree dependable, have been embodied by Admiral Fitzroy in a sort of code of instructions or "forecasts," which have been so very extensively circulated by his praiseworthy zeal, aided by the powerful means at his disposal, that we do not consider it necessary to recapitulate them. They rely mainly on the indications of the barometer and thermometer, together with the observation of the direction and force of the wind at the time and place, and of its immediately previous course; all these particulars being regarded not *per se*, but as in connexion with each other; their indications not being absolute, but relative: so that a rise in the barometer, coupled in one case with a rise, and in another with a fall in the thermometer, may indicate, under given, or, as the case may be, differing circumstances of wind; widely different or even opposite features in the character of the approaching weather. It is to be borne in mind, however, most carefully, that all such indications are to be received as valid (*pro tanto*) only for a very brief interval in advance; and that the "weather-prophet" who ventures his predictions on the great scale, is altogether to be distrusted. A lucky hit may be made. nay, some rude approach to the perception of "a cycle of seasons" may *possibly* be attainable. But no person in his senses would alter his plans of conduct for six months in advance in the most trifling particular, on the faith of any special prediction of a warm or a cold, a wet or a dry, a calm or a stormy summer or winter. Of all the minor or *simply connotative* indications of the coming weather (as distinct from those which connect themselves with our knowledge of causes), the only one in which we place the slightest reliance is that the appearance of "anvil-shaped clouds" is very likely to be speedily followed by a gale of wind.

(7.) The moon is often appealed to as a great indicator of the weather, and especially its changes as taken in conjunction with some existing state of wind or sky. As an attracting body causing an "aërial tide," it has of course *an* effect, but one utterly insignificant as a meteorological cause; and the only effect distinctly connected with its position with regard to the sun which can be reckoned upon with any degree of certainty, is its tendency to clear the sky of cloud, and to produce not only a serene, but a *calm* night, when so near the full as to appear round to the eye—a tendency of which we have assured ourselves by long continued and registered observation. This, however, is more than a "simple connotation." The effect in question, so far as the clearance of the sky is concerned, is traceable to a distinct physical cause, the warmth radiated from its highly heated surface; though why the effect should not continue for several nights after the full, remains problematic.

(8.) Lunar prognostics about the weather may be classed under three several heads,—viz., 1st, Simple connotations of the appearance of halos, coronas, lunar rainbows, and "a watery" moon, as prognostics of wet. No doubt they do indicate the presence of vapour, passing into cloud, in the higher regions of the air (in that of the rainbow, actual rain not far away), and so may be put on a par with the indications which may sometimes be gathered from the behaviour of birds, especially such as fly high, and make long excursions, and which may convey to us some notion of *their* cogitations as to the coming weather; which are perhaps more likely to be right than our own, as founded on a wider range of perception. 2d, Purely arbitrary *laws* or *rules* founded on the hour of the day or night at which the changes of the moon take place. There is (or was a few years ago, for we believe the race is dying out) hardly a small farmer or farm-labourer who had not some faith in certain "weather-tables" in the "Farmer's Almanac," ascribed (we need hardly say falsely) to the late Sir W. Herschel, and which went on this principle. Others, again, pressed into the service the great and recondite names of Apogee and Perigee; and professed to determine the character of the lunation from her proximity at new or full to these mysterious points of her orbit. Both the one and other rule utterly break down when brought to the tests of long-continued and registered experience. Others, again, drew their prognostic for the whole lunation from the character of the weather during the first quarter. Such was the rule said to have been implicitly adhered to by the late Marshal Bugeaud in the planning of any military expedition whose success was likely to be any way dependent on weather:—

> "Primus, secundus, tertius, nullus,
> Quartus, aliquis,
> Quintus, sextus, qualis;
> Tota Luna talis."

(9.) 3dly, A more ambitious form of lunar prediction was that of the late eminent meteorologist (for such, this one crotchet excepted, he certainly was), Luke Howard; who took great account of the moon's declination as influencing the *averages* of rainfall, and of the height of the barometer. Still more so was his weather-cycle of nineteen years, the period of the circulation of the nodes of the moon's orbit; in the course of which the *absolute maximum of north declination* occurs when the ascending node is in the spring equinox, and the moon 90° in advance of the node in her orbit, and that *of south* in the reversed circumstances—the intermediate situations of the node corresponding to the *absolute minima* of each. These situations, according to the declination theory, ought to bring round a periodical increase and diminution in the average rainfalls and barometric heights. Like the others, however, when compared on any extended scale with recorded facts, this results in no establishment of any positive conclusion.

(10.) A small monthly depression in the average temperature arising from the nocturnal radiation consequent on the cloudless state of the sky about the full moon, would seem almost a necessary consequence of that phænomenon.

(11.) The causes by which that "various and mutable thing" which we call THE WEATHER are produced are in themselves few and simple enough; but the physical laws which determine their actions are numerous and complex; and the results, in consequence, so mutually interwoven, and the momentary conditions of their action so dependent on the state of things induced by their previous agency, that it is no wonder it should be next to impossible to trace each specific cause (acting as it has done through all past time) direct to its present effect. Yet from this very complexity results that sort of regulated casualty—that apparently accidental, yet limited departure and excursion on either side from a monotonous medium—that exceeding variety of climate, which renders our globe a fit habitation for such innumerable diversities of incompatible life—and that general equilibrium in each which secures to every species, and to each individual of them all, its due share in the distribution of heat, moisture, and wholesome air:—considerations, these, which are not lost on those who believe that they can trace in nature the operation of motive and design as distinct from a mere necessity arising out of the nature of things and the so-called conservation of *vis viva.*

(12.) Let us take our globe as we find it—revolving on its axis in twenty-four hours; and carried round the sun in an orbit oblique to its equator in a year; which is divided into two somewhat unequal halves (if such an expression may be pardoned) from equinox to equinox by its unequal angular motion in a slightly elliptic orbit; thus giving rise to unequal summers and winters in the two hemispheres:—its surface very unequally divided between land and sea—the land mainly congregated upon one half of it, and that half principally belonging to the northern hemisphere; and so distributed as effectually to bar all free circulation of the ocean in the direction of the diurnal rotation (or round the equator), and allow but a restricted one in that at right angles to it (or across the poles): thus compelling whatever circulation does exist, to take place within three great basins or semi-land-locked areas, and a vast southern expanse into which all the three open; and within each of which a system of circulation is kept up by the action of the winds; its course being determined partly by the sinuosities of their shores, partly by the inequalities of their bottoms, and partly by the rotation of the earth itself.

(13.) We have, besides, to consider the globe as entirely and deeply covered by an atmosphere of mixed gases—highly elastic, very dilatable by heat, and of extreme mobility: expanding itself in virtue of its elasticity out into space, far above the tops of the highest mountains; yet, in virtue of its compressibility, so condensed (comparatively) in its lower strata as that one-third of its total ponderable mass lies within a mile of altitude above the sea-level—nearly one-half within two, and nearly two-thirds within five miles; within which latter limit the whole would be contained, were it everywhere of the same density as on the surface: so that only about one-third of its total mass is free to range, unimpeded by the crests of the highest Himalaya; and not much more than two-fifths can entirely clear the range of the Andes without pressure *à tergo*. In consequence, when driven in the state of WIND over these or other mountain ranges, it is thrown up into vast ripples or waves, which are propagated thenceforward onwards over indefinite areas of land or sea, and become no doubt the origin of a great part of those casual fluctuations of the barometer which give so much trouble to meteorologists.

(14.) This aërial ocean is not of the same temperature throughout, even in the same climate and over the same tract of country. It is everywhere warmer near the ground, colder aloft: and at very great heights a most intense cold always prevails; more intense than that of our severest winters. Hence the snow which covers the summits of lofty mountains even in the hottest climates. This relation between the temperatures existing below and aloft is not subverted by any amount of mutual admixture of the strata, such as internal movements or ascending currents would produce. On the contrary, paradoxical as it may appear, such ascensional movements are the primary cause of this state of things; in consequence of the habitudes of air with respect to heat when compressed or expanded, according to a mode of action well understood by meteorologists, which we need not stop here to explain, as the reader will readily collect it for himself from what follows.

(15.) As the air aloft is colder than below, so also is it *drier*. Every one considers that he knows the distinction between damp and dry air; but many are not aware that all air contains *some* moisture, in the form of transparent invisible vapour; or that in summer and winter on two days, both which would in common parlance be pronounced dry ones, there is more than twice as much moisture present in an equal bulk of air in the summer, as in the winter day. In this state of invisible vapour which water is always assuming (throwing itself off in that form from its surface whenever exposed, and the more copiously the warmer it is), the air is its general recipient and distributor. The mechanism by which it is enabled to do so on the great scale is exceedingly curious. We shall endeavour to exhibit it, as it were in action—not so much with a view to affording a *coup-d'œil* of the whole of meteorology, as with that of rendering in some degree more intelligible than at present it seems to be, that great phænomenon of the November storms, with the mention of which we began this lecture, which has never been satisfactorily explained.

(16.) Looking at our globe as revolving under the warming influence of the sun, whose rays at noon fall on it with little obliquity in tropical regions, while their incidence on those near the poles is always very oblique, and during the half of each year null; it is obvious that its surface must be very unequally warmed. The cook, to use a homely illustration, knows full well that, however good her fire, the two ends of her joint will be under-roasted when the middle is done brown; unless she apply a couple of concave reflectors on her spit to throw some of the lateral heat upon them. As a matter of fact, no one needs to be told that it is so; and that the intertropical regions of the globe are very hot, and the polar, habitually very cold. The average annual temperature at the equator is about 84° Fahr., while in the colder regions near the North Pole it is as low as 5° Fahr., or 27° below freezing. The difference would be much greater were there no sea, or even were the whole surface initially moist soil. Whatever that initial moisture, it would soon *dry off* from the warmer portions, to settle down in snow or hoar frost on the colder; after which the dried portions would grow hotter and hotter. Every one knows what a cooling power there is in the evaporation of water. So long as a vestige of moisture were present, the temperature of the soil could never, at at all events exceed, however it might fall short of, that of boiling water: but when once completely dried off, there would no longer be a limit to the possible increase of temperature; since there would then be no circulation or return of moisture to the part once dried. How this circulation is kept up under the existing circumstances is what we must now explain: and first of all how it happens that in the course of ages the whole ocean has *not* been transferred by this sort of distillatory process from the tropics to the poles; leaving the former dry, and piling the latter with mountainous accumulations of ice. Were the Polar regions of the globe occupied by land instead of by sea, there is every reason to believe that such *would* be the case. As it is, the contrary arrangement prevails, and the Polar snows fall upon these seas or upon their frozen surfaces, and form floating masses of ice, which are partly broken up and drifted away, and partly melted *in situ* by currents of water perpetually streaming in against and beneath them from warmer regions, and thus become restored to the general ocean.

(17.) But what, it will be asked of course, produces these warm currents? And *how* is the water of which that snow consists, and all the rain which falls and feeds the rivers that restore *it* to the sea, raised into the air, and distributed over the world, and thrown down again indiscriminately over all its surface? Common sense assures us that all the rain, &c., which falls from the skies must have originated in the sea, and must (if the present state of things is to endure) find its way back to it. But how is it done? And, in the first place, where are we to look for the motive power? To this the answer presents itself at once. In the sun's heat. Any of our readers who will take the trouble to refer to Lect. II., § 23, will find that the amount of solar heat which actually reaches the surface of our globe would suffice to melt an inch in thickness of ice in two hours thirteen minutes on a surface perpendicularly exposed to it; and from this he will have no difficulty in calculating the depth of water over the whole area of the globe, land and sea, *per annum*, which it would suffice to convert into vapour if wholly expended in so doing. This he will find to amount, as nearly as may be, to nine feet.* Meteorologists, collecting the registers of "rainfall" in all regions of the globe, and comparing and calculating on their indications, have come to the conclusion that taking one region with another, the quantity of water actually precipitated from the air *per annum*, in the forms of rain, hail, snow, and dew, would suffice to cover the whole of its surface to a depth of five feet. Remains *the equivalent* of four feet, expended in warming the soil; which is partly *radiated* away, and partly communicated to the air, thus going to maintain the average temperature, *according to its climatic distribution.* And as solely expended on this last-mentioned object, we have to reckon fully one-third of the sun's total radiation, or one-half of that already accounted for, which is absorbed by the air, or rather by the moisture in it, before reaching the earth. The joint effect of these two portions is, as we have seen, to maintain the air in

* We will make the calculation for him. An inch of ice melted in 2 hours 13 minutes over a great circle of the globe perpendicularly exposed to the sun, corresponds to a quarter of an inch in that time over the whole surface (which is four great circles), or, per annum, to 987·75 inches; or to nine-tenths of this, or 890 inches of water raised 135° Fahr. in temperature; or (taking the initial temperature of the water evaporated on an average at 60° Fahr.) to 108 inches or 9 feet heated 1112° to convert it into steam.

the equatorial region of the earth habitually hotter by about 80° Fahr. on an average of all seasons and hours, than the Polar.

(18.) Hot air under equal barometric pressure is lighter than cold. The equatorial portion of the atmosphere, then, in comparison with the polar, is dilated upwards; the only direction in which the lateral pressure it experiences will permit it to dilate. Hence, the external form of the atmosphere, and of each of its upper strata, instead of conforming in exact parallelism to the spherical* form of the globe on which it reposes, as the laws of equilibrium would require; are unduly elevated, and bulged out, equatorially, into elliptic forms, a state of things inconsistent with repose. The prominent portion rests, in fact, either way, on a slope, and being unsupported laterally, *flows down* on either side—that is, from the equator towards the poles. In so doing, however, it deserts its place, and ceases to contribute by its weight to the total pressure on the equatorial region; while at the same time it goes to add to the weight incumbent on the polar. Thus the hydrostatic equilibrium of pressure is subverted, and air is pressed inwards towards the equator from the poles below, to make good the efflux aloft. A circulation is established in each hemisphere by inferior currents of air running in on both sides towards the equator and superior ones setting outwards, all around the globe, from the equator towards the poles. Both these, were the earth at rest, would

* We neglect the spheroidal form of the earth, which in meteorology is never worth considering.

follow the directions of meridians, but are converted by its rotation on its axis and the gradual diminution of the rotatory velocity in advancing from the equator to the poles, into relatively oblique currents. The upper or outward follow precisely the reverse direction to the lower or inward, and being drawn downwards, and are ultimately brought down to the sea level, in their approach towards the poles to supply the void which would otherwise be left by the withdrawal of air below.* They thus become surface winds, prevalent in the regions beyond the tropics from about 30° of latitude either way.

(19.) In the view thus taken of the great and permanent system of winds known as the "trades" and "anti-trades," it will be observed that we have been careful to regard them as resulting not so much from the immediate (and diurnally intermittent) action of the sun, as in a certain established gradation of climatic temperature, the result of its action on the whole earth's surface continued through successive ages. Were the sun extinguished, the system of the trade winds would continue to subsist, though with diminishing intensity, so long as the equator continued in any degree warmer than the poles.

(20.) By the action of the trade winds which occupy

* Those of our readers who are not already familiar with the nature of this transformation, and who would wish to follow it out more closely, are referred (as well as for every other matter of detail in similar cases, precluded by our limits) to the article Meteorology, in the "Encyclopædia Britannica," or to the same article published separately by (A. & C. Black, Edinburgh) the editors of that work.

the intertropical region, and a little more, and which, though differing as to north and south, conspire in their general easterly character, the surface of the equatorial ocean is driven westward, and directed full against the two great barriers (the west coasts of America and Asia), and divided northward and southward into streams or currents which in their progress, after issuing from tropical latitudes, receive a direction, by reason of the rotation of the earth, corresponding to that of the anti-trade winds. These also, beginning about the same latitudes to descend to the sea level and strike on the ocean, aid their further progress, and carry them, or portions of them, far northward and southward into the Polar Seas, there to perform the work above assigned to them of melting the ice, and so keeping up the total amount of the ocean-water; besides mitigating, to a great extent, the severity of the cold on the coasts in high latitudes on which they strike; of which we have a familiar example in the warming influence of the celebrated Gulf-stream.

(21.) The steady and equalized agency by which the great system of the permanent winds and oceanic currents is kept up, which we have just described, contrasts itself strongly with the violent and, as it may almost in comparison be called, impulsive action of the sun on and around the point of the globe over which, for the moment, it happens to be vertical; and which corresponds to that portion of the solar energy which is directly employed in producing evaporation. The nature of this process we have now to explain.

(22.) When water is converted into invisible vapour, it occupies between sixteen and seventeen hundred times its original volume, and becomes much lighter than air—as light, indeed, as the ordinary coal gas with which balloons are filled, so that if enclosed in a similar envelope it would rise in the air like a balloon. Being free, however, it mixes with the air, and *that* not merely by a simple chance-medley confusion, but by a peculiar self-diffusive energy arising from its inherent elasticity; by which the particles of every one species of gas or vapour struggle to interpenetrate, and needle, as it were, their way among those of every other. These latter oppose to them no *elastic pressure*, but that simple resistance to jostling which an inert body of any other kind might do, —which feathers, for instance, might oppose to air, introduced and struggling to diffuse itself among them. Of course they will be pushed from their places in the struggle, both laterally and vertically, and thus arises over the whole region in which the vapour is in course of production, a pressure on the air both outwards and upwards. The former, however, cannot be effective in removing air bodily to any great distance horizontally, for the simple reason that to do so it would have to *shove aside* the whole surrounding aërial atmosphere, and to crowd it upon that which is beyond. while there is room in a vertical direction for an indefinite removal, and the upward pressure is also aided by the lightness of the up-struggling vapour, which therefore rises rapidly—*not without dragging up with it a great deal of air.* The consequence is to establish, immediately under the sun, at whatever part of the globe it happens to be vertical, and at which there is a supply of moisture, and for a very large space around it; what may be likened to a vast up-surging fountain of air and vapour throwing itself up with an impetus; breaking up and bulging outwards the immediately incumbent aërial strata very far above their natural levels; and introducing at the same time into the air a great quantity of vapour, as well as withdrawing, *by direct transfer*, from the lower atmosphere, a great deal of air; which of course has to be supplied by in-draft along the surface of the earth.

(23.) The process now described, is in a great many of its features similar to that gentler one previously stated: and as it always takes place at some point or other of the intertropical region, it conspires with and locally exaggerates its result so far as the transfer and circulation of air and the production of winds is concerned. As regards the vapour, a large portion is very speedily deprived of its elasticity and ascensional power, and reduced to the state of visible cloud, collecting and descending in rain. This is a consequence partly of its arrival in a colder region, but mainly of the property which all gases and all vapours alike possess, of absorbing and rendering latent a large quantity of heat as they expand in volume, and so becoming, *ipso facto*, colder. Both the air and the vapour *do* so expand as they rise, by reason of the diminution of pressure they experience. The air indeed retains its elastic state *as air*, however cold it may become; and therefore merely takes its place in its new situation *as very cold air*, without further tendency to rise. But the vapour so chilled loses its vaporous state, and condenses in the manner above stated; leaving only so much uncondensed *as can remain vaporous under that temperature and pressure.* This is the origin of those continual and violent tropical rains which always accompany the vertical sun, and its near neighbourhood, and of which we feel the influence, though slightly, in our wet Julys. The vapour being thus arrested in its upward progress, the whole of the evaporatory process we have just described, however tumultuous in its origin, is confined to what may be considered comparatively the lower strata of the atmosphere. But these become in this manner saturated with moisture; and when carried into the general circulation, convey it either as cloud or as invisible vapour to the farthest regions of the earth.

(24.) Besides the evaporation produced by the direct action of the sun, a vast amount of moisture is taken up by the air immediately from the sea and land over which it passes in its indraft towards the Equator as a trade-wind. Coming from a colder region to a warmer, and acquiring heat as it advances, its capacity for receiving and retaining moisture in an invisible state is continually increasing; and hence, even during the absence of the sun in the night hours, it is constantly absorbing moisture; which it carries along with it, and delivers, as a contribution of its own collecting, into the general ascending mass, to be handed over in the returning upper current into the circulation. Hence it arises that the regions of the earth habitually swept by the trade-winds abound in sandy deserts and arid wastes. When, however, in the progress of that circulation, it descends again to the earth, and becomes a surface-wind (assuming the character of an "anti-trade"), it finds itself in precisely reversed circumstances. It is now travelling from a warmer to a colder region. Saturated with moisture in the warmer, and parting with the heat which alone enabled it to retain it, its vapour condenses. Clouds already formed thicken, and descend in rain, and fresh ones are continually forming, to fall in snow at a further stage of its progress; till all the superfluous moisture is thus successively drained off, and it is prepared to re-assume, while starting on a fresh circuit, the character of a drying wind.

(25.) We have here the origin of that generally observed difference of character between our two most prevalent winds—the S.W. and the N.E. The former is our "anti-trade," that which from our geographical position we are chiefly entitled to expect, and which, in point of fact, is of far the most frequent occurrence. Its prevailing characters are warmth, moisture, cloud and rain, as well as persistence and strength. In the former of these characters it is strongly reinforced by the circumstance of its accompanying across the Atlantic the Gulf-stream, which, in fact, it helps to drift upon our western coasts, and which, retaining a considerable amount of the equatorial heat, sends up along its whole course a copious supply of vapour, in addition to that with which the air above it is already loaded: and this it is which gives to our west coasts, and to that of Ireland, their moist and rainy climate—double, and more than double, the amount of rain falling annually on the coasts exposed to its full influence, as compared with the eastern coast; which it does not reach until drained of its excess of humidity.

(26.) The characters of our North-east winds (for such as are in common parlance called Easterly winds are almost always such) are the reverse of these in every particular. They are cold, dry, and hence often spoken of as *cutting*, from their parching effect on the skin; and, as a natural consequence, for the most part accompanied with a clear sky. They are seldom of very long continuance, and may be regarded rather as casual winds, except in the spring; when the advance of the sun to the north of the equator begins to call into action a northern indraft—to push to the northward the limit of the north-east trades, *and to unsettle by its intrusion the line of demarcation between the wind-zones which its long continuance in extreme south latitude, near the winter solstice, had allowed to take up, and rest in, its extreme southernmost position.* To this opposition of characters we may add, that the South-west wind is generally accompanied with a lower, and that of the North-east with a higher than average barometric pressure; a connexion partially, but not entirely, accounted for by the lightness of warm and moist air as compared with cold and dry; and which is the origin of those indications of the weather (*fair, settled fair, rain, much rain, &c., &c.*) which we find inscribed opposite to the divisions of the scale of inches in our ordinary barometers. When the North-east wind brings snow, as it very frequently does, it is not by the precipitation of its own moisture; but by its intrusion as a cold wind into a warmer atmosphere charged with moisture, and ready to deposit it under any cooling influence.

(27.) Complementary to the phænomenon just mentioned of a tendency to North-easterly wind in the spring, *i.e.*, to the production of a lull or temporary intermittence in the regular South-west current, and the substitution for it of its opposite; may be considered that aggravation of its intensity which takes place subsequent to the autumnal equinox; exaggerated, however, and thrown later into the season, viz., into November, by the conspiring action of several distinct causes, which we will now proceed to explain.

(28.) As the sun in its annual course traverses the northern and southern halves of the ecliptic, it creates summer in the one hemisphere, simultaneously with winter in the other; and the balance of aërial expansion and aqueous evaporation is alternately struck in favour of each. As a necessary consequence, a large amount both of air and of aqueous vapour carrying air along with it, is alternately driven over from one hemisphere to the other. The only course which the elements so transferred can pursue, is by passing in the higher regions of the atmosphere across that medial line where the two superior out-flowing currents separate on their courses towards either pole—in other words, by joining with, and reinforcing the "anti-trade" current on that side of the equator *towards* which they are propelled.

Now this cause of reinforcement cannot begin to be felt until the sun, having passed the equinoctial, has advanced considerably towards the other solstice. In the case of the northern anti-trade, the effect in question is rendered still more sensible by the great preponderance of sea in the southern hemisphere as compared with the northern ; and the much greater quantity of vapour raised by the summer sun on that side of the equator. And besides all this, it will be remembered that all the *air* which had been dragged across the equator into the southern hemisphere by transferred vapour during the continuance of our northern summer, and there as it were imprisoned, is now released; and returns, necessarily by the same course, and contributes to reinforce the northern anti-trades.

(29.) There is a special cause, too, arising from the geographical position of Britain and north-west Europe, in relation to the South American continent, which is probably not uninfluential in producing or aggravating this disturbance. If we trace on a map the course of the wind which reaches our island from the south-west, we shall find that it has its origin on the coast of Guiana, between the fiftieth and sixtieth degrees of west longitude. This also is nearly about the point where the medial line between the north and south trades, in its average position, intersects the South American coast. Here the South American continent is comparatively narrow, but south of this it expands in longitude, and between the fifth and fifteenth degrees of south latitude has an average breadth of between 30° and 40°.

(30.) In the view we have taken of the production of the trades, the immediate verticality of the sun acts as a disturbing force. In its passage from solstice to solstice it causes an annual fluctuation or oscillation to and fro of this medial line, and of the northern and southern limits of the wind-zones; which, where those limits cross the ocean, is but of moderate amount, because the medium temperature of the intertropical seas varies but little with the seasons. But where they cross extensive tracts of land, their oscillations to and fro may become very considerable, owing to the high temperature which the land is capable of acquiring. Now in this case, so soon as the autumnal equinox is passed, the vertical sun enters on the full breadth of this vast continental tract; and commences throwing up torrents of vapour and intensely heated air, the latter being far in excess of what it would be over an equal area of sea; while at the same time, owing to the sun's then rapid motion in declination, the limits of the wind-zones retreat southward, and their regularity is disturbed and broken; which cannot but give rise to great temporary confusion and disturbance in the winds themselves. As to the "atmospheric wave" which recurs periodically at this season, it results most probably from the operation of the more general of the causes above mentioned, by which a large amount of extraneous air and vapour is thrown into the atmosphere *of the North Pacific;* causing the south-west wind of that ocean to sweep with increased force up the western slope of that vast range of lofty mountains which fringes the North American continent; and to be thrown up along the whole length of that range into a broad swell, propagated onwards as a wave across America and the North Atlantic into Europe. No merely local action, and no casual conjunction of circumstances, can be considered competent to produce so extensive and so regularly recurring an effect. A mere inspection of Admiral Fitzroy's interesting compendium of the state of the barometer, &c., &c., over the area occupied by our island and the neighbouring continental coasts, as recorded from day to day in the *Times*, will suffice to satisfy any one of its occurrence in former years, and to show that its character has been (so far at least, Nov. 21) fully maintained in the present (1864).

(31.) If we are ever to make any material progress in the prediction of the weather beyond "forecasts" of a few hours, or it may be a whole day in advance, it can only be by the continued study of such of its phases as recur periodically, or of such as manifest a *periodicity of event*, as distinct from that of times and seasons, with a view to connecting them with their efficient physical causes. Of this latter description we have an example of one, and of its successful reduction under the domain of philosophical reasoning, in the law of the rotation of the winds. That the winds in their changes, in a general way, "follow the sun," *i.e.*, have a tendency to veer in the same direction round the compass card with the sun's apparent diurnal course in the heavens (from east round by south, west, and north in the northern hemisphere, and reversely in the southern), in continual succession back to the original point—has been surmised from very early times; but until lately, rather as a matter of occasional remark, agreeing on the whole with the general impressions of casual observers, than as a meteorological law of universal applicability. As such, however, it has now taken its place among ascertained facts; verified by the registered movements of the wind-vane at every station where continuous observation is made; and connected by the researches of M. Dove with that great fact which underlies so many other phænomena—the rotation of the earth on its axis.* Nothing apparently can be more capricious than the shifting and veering of a weathercock on a gusty day, and to any one who watches its leaps to and fro for a few hours, it may well be a matter of surprise to be told that with anything like a fair exposure, the preponderance of its movement is sure to be in one direction—if not in a week or two, at all events on the long average, and in a great majority of cases before the expiration of a month. Thus it appears from the record kept at the Observatory at Greenwich, in which every change of the wind's direction is noted by a piece of mechanism attached to the vane and traced on a table by a pencil—that in the thirteen years elapsed from the beginning of 1849 to the end of 1861, the vane made 166 complete revolutions more in the direction

* For the reasoning by which this connexion is made, and for the mode in which any casual advance and retreat of a body of air over an extensive but limited tract of country is transformed by this cause into a relative gyration, the reader is referred to the works already cited in a former note.

above indicated than in the opposite, on a comparison of the sums of all its angular movements either way—or on an average, nearly thirteen revolutions *per annum.* In all this interval, two years only, 1853 and 1860, gave a contrary result, and that only to the total amount of two revolutions in excess the wrong way in each. And of these the year 1860 was in many points an abnormal one in respect of stormy weather. Nothing can convey a better idea of the disappointment to which all meteorological predictions, even though founded on just principles, and supported by extensive inductions, are liable, than this example. Still there remains a decided *balance of probabilities* in favour of a change of wind occurring in this rather than in a contrary direction on any specified occasion. A continuous circuit round the horizon in the contrary direction would certainly be in a high degree improbable.

(32.) On the other hand we have an instance of the failure of a distinctly periodical cause (as to all appearance it would seem fairly entitled to be considered *à priori*), to exhibit itself in any cognizable periodical effect on the seasons, in that curious recurrence of a spotted state of the sun's surface which takes place every eleven years (see Lecture II., § 36). Looking to the sun as the great source of all meteorological action, it might most reasonably be expected that such indications of an activity of *some sort* going on in its very photosphere—in the actual visible laboratory of its light and heat—would correspond to some difference in its supply of both; which, recurring periodically at stated intervals, *must*, one wo 'd think, manifest itself in some effect or other on oui weather and climates. Such, however, does not yet appear to be the case. The most obvious consequence would seem to be a periodical return of *hot* and *cold* years, which, however, the average registered temperatures of successive years in different places have not borne out. Yet, after all, it is possible that meteorologists may here have been on a wrong scent, and that increased emission of heat from the sun may make itself felt, not so much in any material increase of the average annual temperature, as in an increased generation of vapour from the ocean; in a much more copious and immediate rainfall in the equatorial regions of the globe, and in a sensible increase of it over the whole earth's surface: but especially in a more cloudy state of the general atmosphere, consequent on the introduction of a larger amount of vapour into it; and in an increased tendency to atmospheric disturbance and barometric fluctuation. No one who has watched with disappointment the rapid upcast of cloud on a calm morning commencing with unclouded sunshine, which blots the prospect of a glorious summer day, and who has seen the same change take place day after day, often for weeks in succession, can have failed to be struck by that self-induced interposition of a veil between the sun and the earth's surface which mitigates the ardour of his beams and tempers them to the requirements of animal and vegetable life. The increased heat, or by far the greater part of it, may be expended in *re-evaporating the upper surface of this very cloud*, and, by so doing, simply extending the limits of the vaporous atmosphere and maintaining the higher regions of the air in a state of increased humidity. And this is the way in which we conceive it possible the planets Venus and Mercury (as we have before hinted in our Lecture on the Sun, but without further explanation) may be maintained at a degree of superficial temperature not incompatible with even terrestrial forms of life. Their climate, to be sure, would have little to recommend it to our tastes; as it would probably afford small relief from a perpetual succession of cloudy days and rainy nights.

(33.) Is it in any degree within the power of man to *alter* the weather? A strange question, it may seem at first sight to propose! but by no means so absurd a one as it may appear. The total amount of annual rainfall over any district, is an element of its weather and its climate of the last importance; and when we look over the registers of rainfalls which are now so assiduously kept in almost every part of England and other civilized countries, it is impossible not to be struck by its very great local diversity; even in neighbouring places, whose general similarity of situation as regards wind-exposure and surface *configuration* would seem to preclude any material difference on an average of years in their reception of rain; if really indifferent in its *choice where to fall.* There is evidently something distinct from mere local *situation* which determines this element of climate; and we must look for it in the nature of the surface of the districts, and its relations to heat and moisture—relations which the operations of man on the soil itself, and his selection of the kind of vegetation with which it shall be habitually clothed, place to a great extent within his power. It is chiefly in his clearance or allowance of arborescent vegetation, and in his artificial drainage of the soil over extensive districts for agricultural purposes, that his influence on these relations is perceptible. The total rainfall, and (which is perhaps as regards weather and climate of even more importance) the *frequency* of showers on an extensive well-wooded tract, or one entirely covered by forests, ought, on every theoretical view of the causes which determine rain, to be greater than on the same tract denuded of trees. The foliage of trees defends the soil beneath and around them from the sun's direct rays, and disperses their heat in the air, to be carried away by winds, and thus prevents the ground from becoming heated in the summer; while, on the other hand, a heated surface-soil reacts by its radiation on the clouds as they pass over it, and thus prevents many a refreshing shower, which they would otherwise deposit, or disperses them altogether. So again of drainage :—by carrying away rapidly the surface-water down to the rivulet, and so hurrying it away to the ocean, it not only cuts off a great deal of the supply of local evaporation, which is a material element in the amount of rainfall, but by causing the surface to dry more rapidly under the sun's influence, it allows it also to become more heated; and so to conspire with the action of the denudation of trees to prevent rain. Evidence is not wanting to corroborate this *à priori* view of the matter. The rainfall over large regions of North America is said to be gradually diminishing, and the climate otherwise altering, in consequence of the clearance of the forests; while, on the other hand, under the beneficent influence of a largely increased cultivation of the palm in Egypt, rain is annually becoming more frequent. Lakes are cited in what was formerly Spanish America whose water supply (derived of course from atmospheric sources) had been so largely diminished, owing to the denudation of the country under the Spanish regime, as to contract their area, and leave large tracts of their shores dry; which, now that the vegetation is again restored, are once more covered by their waters. Even in our own southern counties complaints are beginning to be heard of a diminution of water supply, partly, it is said, owing to gradually decreasing rainfall from the universal clearance of timber,* though chiefly perhaps attributable to robbing the springs of their supply by draining—a practice beneficial no doubt to agriculture, if used with caution and in moderation, but of which *the consequences, if carried to excess, may ere long be very severely felt, in rendering large tracts of*

* On the other hand, forests, owing to the immense evaporation from their foliage which must be supplied from the soil beneath, have a direct tendency to *drain that soil upwards*, and so throw its moisture into the air. This has been well pointed out and strongly insisted on by M. le Marechal Vaillant, in "Les Mondes," T. 8, p. 674. As a matter of fact, it seems pretty distinctly proved by the collection of data laboriously accumulated by Mr Symonds—that the annual average rainfall *is* decreasing over the whole of the British Isles, and more especially along a line running nearly S.W.—N.E. from Cornwall to the Wash. (Symond's Report of British Association, 1865.)

country uninhabitable in summer from mere want of water.

(34.) To return to our prognostics. We would strongly recommend any of our readers whose occupations lead them to attend to the " signs of the weather," and who, from hearing a particular weather adage often repeated, and from noticing themselves a few remarkable instances of its verification, have " begun to put faith in it," to commence keeping a note-book, and to set down without bias all the instances which occur to them of the recognized antecedent, and the occurrence or non-occurrence of the expected consequent, not omitting also to set down the cases in which it is left undecided; and after so collecting a considerable number of instances (not less than a hundred), proceed to form his judgment on a fair comparison of the favourable, the unfavourable, and the undecided cases: remembering always that the *absence of a majority one way or the other would be in itself an improbability*, and that, therefore, to have any weight, the majority should be a very decided one, and *that* not only in itself, but in reference to the neutral instances. We are all involuntarily much more strongly impressed by the fulfilment than by the failure of a prediction, and it is only when thus placing ourselves face to face with fact and experience, that we can fully divest ourselves of this bias. Any one before whose eyes these pages may pass, for instance, who may feel disposed to give our *dictum* respecting the clearance of the sky under the influence of the full moon (we will not say through a hundred lunations, but throughout the year 1864) a fair trial of the kind, should record the state of the sky as to cloud on three successive nights of each lunation—that on which the full moon occurs, and those immediately preceding and succeeding it, from an hour before the rise of the moon, and thence hourly to as late an hour of the night as his usual habits will allow—noting the strength and direction of the wind, and accompanying his memoranda in every instance with a note of the day and hour at which the observation it refers to was made.

LECTURE V.

CELESTIAL MEASURINGS AND WEIGHINGS.

"Omnia in numero, pondere, et mensurâ."

HERE are epochs in the history of every great operation, and in the course of every undertaking, to which the co-operations of successive generations of men have contributed (especially such as have received their increments at various and remote periods of history); when it becomes desirable to pause for a while, and, as it were, to take stock; to review the progress made and estimate the amount of work done: not so much for complacency, as for the purpose of forming a judgment of the efficiency of the methods resorted to, to do it; and to lead us to inquire how they may yet be improved, if such improvement be possible; to accelerate the furtherance of the object; or to ensure the ultimate perfection of its attainment. In scientific, no less than in material or social undertakings, such pauses and *resumés* are eminently useful; and are sometimes forced on our consideration by a conjuncture of circumstances, which almost of necessity obliges us to take a *coup-d'œil* of the whole subject, and make up our minds, not only as to the validity of what is done, but of the manner in which it has been done; the methods employed; the direction in which we are henceforth to proceed, and the probability of further progress.

(2.) The subject to which this lecture is devoted affords an instance of a conjuncture of this kind. We have already had occasion incidentally, in Lect. III. § 9, to call attention to the change which it has been found necessary to make (at present of a provisional rather than a definitive character) in our estimate of the distance of the sun—a change, implying of course the necessity of a proportionate alteration in all those statements of the dimensions of our system, such as the diameters of the planetary orbits; of the sun and the planets themselves; and the distances of their satellites from the primary, and even the estimate of the masses of all these bodies and the dimensions of the cometary orbits: all those elements, in short, which assume directly or indirectly the mean distance of the sun as their unit of scale. There is reason to believe, too, that the distance of the moon (our knowledge of which does *not* assume that of the sun as known) has been somewhat misestimated, and that an alteration (though not nearly to so great a proportional extent), bringing our nearest celestial neighbour into somewhat closer proximity than heretofore supposed, is required.

(3.) The dimensions and figure of the earth itself too,

as concluded from the immense series of great Trigonometrical Surveys carried on now during nearly two centuries, have quite recently, and in two distinct and independent quarters,* undergone a fresh, and most searching and elaborate inquiry. And the conclusion from both is, that our knowledge on this point is not likely to be improved in any material degree by any further operations of the kind; at least until the time, probably yet far distant, when the Australian Continent shall have become easily and conveniently traversable from North to South, and when the wastes of Patagonia and Terra del Fuego shall afford to future geodesists the opportunity of winning a hard-earned distinction. Till then (and most probably then also), we must rest satisfied with the conclusions arrived at,—conclusions, be it observed, which have disclosed a numerical relation of singular simplicity between our British unit of measure and the length of the earth's polar axis.

(4.) Moreover, in ignorance probably of this last-mentioned fact, and therefore with too gratuitous a contempt for our national and time-honoured standards, and too hasty a preference for the apparently more scientifically, and certainly more symmetrically, constructed system of our continental neighbours, an agitation is and has for some time been going on, headed by persons of considerable influence, and strongly, no doubt, though we think unduly, impressed with the advantage of the change; with the object of abolishing *in toto* our British

* By Gen. de Schubert (Mem. Imp. Acad. Petersburg, 1859), and Capt. A. R. Clarke, R.E. (Mem. R.A.S., 1860).

system of weights and measures, and introducing in its stead the French metrical system. A bill was introduced in the session of 1863 into Parliament with this avowed object: and though withdrawn, after passing the second reading, has been reintroduced in the present (1864), and reached the same stage, with every prospect of being passed.* It is true that the change *immediately* proposed is permissive, not compulsory: but there can be no doubt that the attempt, if successful, will be followed up at no distant period by the introduction of a compulsory measure; one whose effect on the habits, feelings, and interests of nine hundred and ninety-nine out of a thousand persons in the whole community, this is not the proper place to dilate upon.

(5.) As civilization extends, wants and desires of a higher order than material gratifications arise; and among them that of extending knowledge *for the sake of knowing*, the craving after a larger grasp, a clearer insight, a more complete conception in all its relations of the wondrous universe of which we form a part. Such desires, when accompanied with the means of their gratification, are included by the author of a recent work of much interest on the subject of wealth (under the somewhat inappropriate title of Plutology†), among those

* It has passed, and is now the law of the land. So far there is no actual harm done, beyond unsettling opinions and creating uneasiness; but we trust the common sense of the nation will repudiate any attempt to carry out to its designed completion a measure so thoroughly retrograde.

† By Professor Hearn, of Melbourne University, Australia. The title ought to have been Aphnology. Aphnos, or Aphenos (αφνος,

sources of positive enjoyment which are not less real because they are intellectual, or less valuable because they cannot be appropriated or bartered in exchange: but which yet cannot be attained by mere intellectual aspiration or effort; but require for their production and dissemination appliances and means of a refined character, and combinations of a recondite kind; such as only an advanced stage of material as well as intellectual progress can furnish. Such a piece of intellectual wealth is the solution of that great enigma (such, at least, in all former time) of the distance of the stars,—a problem which has yielded, at length, to the delicacy and refinement of astronomical observation, during the lapse of the last thirty years; combined with and acting through the marvels of mechanical skill and workmanship which are now obtainable. That distance is now no longer the hazy and absolutely indefinite matter of conjecture which it was (to go no farther back) in the time of Newton, or even in the middle of the last century. Of some, at least, of them it can be said with every reasonable assurance of probability, that their distance is known within an eighth or a ninth part of the truth, one way or the other; and of several, that we can arrange them in order of distance, nearer and more remote, with little or no presumption of mistake. A stepping-stone is thus laid for another upward struggle towards the infinite—to the nebulæ, the remotest objects of which we have any know-

αφενος, Gr.), expresses *wealth* in its largest sense of general abundance and well-being. Ploutos (πλουτος, Plutus), *riches*, in the more restricted sense of the precious metals, or, at the utmost, of exchangeable value.

ledge: though the stride is here too vast, as it may seem, for the limited faculties of man ever to take. Nor can it be said that man is dwarfed and humiliated by such attempts:—that they teach him nothing but his own insignificance, and so are the reverse of ennobling. The greatness of nature is not synonymous with the littleness of man. That only is little which cannot rise to great conceptions.

(6.) Enough, perhaps, has been said to justify an attempt to lay before our readers the actual state of our positive knowledge of these subjects; of the methods by which it has been attained; and of the history of their development; and to give them a distinct conception of the several links which connect the *British standard inch* with the distances of the fixed stars, and of the sort of intermediate units we have to deal with in the inquiry. It fortunately happens that we shall not need for our purpose any resort to abstruse considerations, or have occasion for any illustrations which are not of the simplest kind.

(7.) In every country having the slightest pretensions to civilization there is preserved, with more or less care, some rod, bar, ruler, or other *standard*, the material incorporation of the national idea of the most ordinary unit of length; and its representative, when it is required to test the correctness of one in vulgar use, or to multiply and disseminate copies of itself for purposes of ordinary mensuration. Thus, among the Jews, we find (Exodus xxx. 2) the cubit* identified as equal to either of the

* This cubit is presumed to be identical with the Egyptian cubit, still preserved, of the Nilometer.

sides, or to half the height of the altar of incense. And the same is true of similar representatives of the most commonly received units of weight and measure of capacity. What the original type might be, which such standard professed to represent, matters little. The inhabitants of a nation might agree to use for their unit of length the foot of one of their ancient heroes, or the hundredth part of the height of their principal church, or the hundred thousandth part of the extreme breadth of their country from sea to sea. But as these objects could never be appealed to for the settlement of any practical dispute between man and man, or to convict the user of any fraudulent measure, a material and producible object must exist in some safe custody, carefully preserved, or safe in its received sanctity, from damage; and authoritatively declared, and generally believed to be, rigorously equal in length to its prototype; and to have been, at some period, however remote, ascertained to be so by some appropriate process of comparison; or, at all events, by the exact copying of some former and lost standard so compared. And from the moment of such authoritative declaration, the length of this material representative necessarily becomes the *real* and *legal unit of length.* The hero may turn out, on a close and irreverent scrutiny of history, to have been a purely mythical personage; the church may have been consumed by fire; the breadth of the land diminished by the encroachments of the sea: but so long as the standard remains uninjured by rough usage, and secured from loss by a sufficient multiplication of authentic copies, its practical utility is unimpaired by such mishaps; and should it be really damaged or lost, public opinion readily transfers the same reverence to its legitimate successor.

(8.) The history of our existing "Imperial" standard is not quite so simple. It is the successor of one destroyed by fire in 1834: not, however, being copied or even having been immediately compared with its predecessor; but recovered by the evidence of an assemblage of other standards which had, at various times, been compared with that and with each other. And that again had been derived, not by direct copying and exact equalizing with *its* predecessor the then "reputed Exchequer Standard," but by a somewhat similar process, from all the best evidence that could be procured of a former state of things. The ultimate prototype is either to be referred to the age of Henry I., who is said "to have settled the yard by the length of his own arm," or to the more ancient foot of twelve inches, "each the length of three barleycorns from the middle of the ear, laid end to end." The point is not of the slightest importance, now that we are assured from the number and exactness of the copies taken; their wide distribution; and the precautions taken to ensure their preservation; that it is scarcely in the power of accident to deprive us of a perfectly "legitimate" successor in the sense in which we have above used the term.

(9.) To measure lengths of many miles (to say nothing of the breadth of a country or of a kingdom), by the simple repetition and laying end to end of yard measures (supposed exactly equal), would not only be intolerably tedious but impracticable, except on a carefully-levelled plain free from all obstructions. Nevertheless, when the object is to measure any large tract of country, or to construct a chart of a territory by what is called a "*Trigonometrical Survey*," it is indispensable to lay down and measure, no matter at what cost of time and labour, some one such very long line, as a "*base line;*" and to mark its two extremities in some very distinct and permanent manner: so that their linear distance (a large multiple of the original standard unit) shall not only be exactly known, but shall be capable of being appealed to for all future time, or at least till the whole work is completed, as a new and larger unit, "*the length of the base*," to which all other distances in the survey are temporarily referred. These, being subsequently reduced by calculation to multiples and fractions of the original unit, all the dimensions of the territory become finally known in yards, feet, and inches. For the purpose of measuring such a base, the ground must be cleared and reduced to perfect horizontality (or any slight inclination exactly taken account of), and the intended base line *allineated* by placing a telescope a little beyond one of its proposed extremities, so as to command them both, and as it were to fore-shorten its whole length into one point, the intersection of two wires in its focus. Anything seen in the telescope to the right or left of this point, or above or below it, is out of the line.

(10.) Whenever lengths are to be added by the repetition of one and the same unit, there is always a possibility of error arising from imperfect juxta-positions

And the oftener the unit is repeated (when it once becomes wearisome), the greater is the difficulty of keeping up the necessary attention, and the greater therefore the amount of error to be feared in each case. To diminish this source of accumulating error (besides the saving of time), it is desirable to diminish the number and increase the nicety of these juxtapositions. Hence the utility and convenience of creating an intermediate unit or set of such units or "Base-measuring bars," and of devising some means of juxtaposing or laying them end to end, without the derangement of one by the small shock arising from the contact (however delicately performed) of its successor. These bars should not be so long as to prevent their being conveniently manageable, yet long enough to diminish greatly the requisite number of *their* repetitions. The bars now actually used for this purpose are miracles of ingenious contrivance and delicate workmanship. They are *self-compensating for changes of temperature;* that is to say, the two fine dots which mark the two extremities of the measure remain exactly at the same distance from each other whatever be the temperature of the bars, which are compound ones of two differently expansible metals combined on a principle devised by the late Lieutenant Drummond. And their repetition is performed, not by driving the end of one against the other, or by laying dot against dot, but by *focussing* a detached microscope on the more advanced dot, removing the bar and bringing the other dot under the microscope to occupy the exact position in the centre of its field (marked by a cross wire) which its predecessor did.* Thus the bar has been moved forward by its exact length *in the air*, as it were, without touching anything. Arrived at the end of the base, a dot is made and adjusted under the terminal microscope on a gold or platina plate let into a solid block of stone already prepared—the starting point having been a similar one similarly fixed at the other end.

(11.) The base measured, the "*Triangulation*" commences. This is founded on the universally known fact that when two angles of a triangle are known, a knowledge of the length of the side between them leads by exact rules of calculation to that of the other two; accordingly, at the two extremities of the base, and centrally over the dots which mark them, are placed delicately divided instruments called theodolites, competent to the measurement of angles to an extreme nicety. The telescopes of these being pointed so as to look down the throats of each other, it is clear that both must be directed along the base line, and if then turned on some one object at a distance considerably greater from either than they are from each other, that object becomes the summit of a triangle, the inclinations of whose sides to the base is measured. Its distance from either end of the base then can be calculated. Thenceforward either of those sides becomes available as a new and longer base. And thus the survey may go on, throwing out

* In actual practice the procedure is a little more complex, but the principle is the same; and it is only intended here to convey to the uninitiated a general notion of the sort of niceties which have to be attended to.

new triangles on all sides, of larger and larger dimensions; till the whole surface of a kingdom or a continent becomes covered with a network of them, all whose angular points are precisely determined. The strides so taken, moderate at first, become gigantic at last: steeples, towers, obelisks, mountain cairns, and snowy peaks, becoming in turn the stepping-stones for further progress; the distances being only limited by the range of distinct visibility of objects through the haze of the atmosphere. Even this is extended by artificial means —by Bengal lights at night and by the use of the "heliotrope," a contrivance of the celebrated Gauss for reflect- a strong sunbeam from station to station; by the use of which, stations 90 or 100 miles distant have been brought into direct connexion.

(12.) If the earth's surface were a plane such a process might be continued *ad infinitum.* The general *roundness* of the earth, however, has been recognized as a fact from very early ages; and indeed it is scarcely possible for any thinking person, with ever so slight an acquaintance with the most elementary geometry, not to be aware of it. It was not, however, till about some three centuries before our era, that something like just notions of its actual size were entertained: unless we admit (an opinion which has found strenuous defenders) that this important datum was already *exactly* known to the ancient Egyptians 1800 years previously;* a fact, *if true*, all memory

* The height of the great pyramid from base to apex is (*was*) contained *exactly* 270,000 times in the circumference of *some one* diametrical section of the earth.

of which had perished in the lapse of that interval. The Chaldean astronomers at the later epoch above mentioned are reported to have arrived at an estimate not very remote from the truth. But the first estimate which has been handed down to us, accompanied with a statement of the process by which it was arrived at, is that of Eratosthenes (B.C. 250); who, measuring the shadow cast by a vertical rod on the day of the summer solstice at Alexandria, and coupling it with the fact reported to him, that at Syene in Upper Egypt on the same day, the bottom of a well received the full sunshine, concluded a difference of latitude between the two places, equal to one 50th part of the circumference of a meridian. Hence, imagining the two places to lie pretty nearly north and south of one another, he concluded the circumference of the earth to be fifty times the distance from Alexandria to Syene, which on the most probable interpretation of his estimate of that distance in the itinerary measures of the time, affords an approximation to what we now know to be the truth, by no means contemptible; falling within about a sixth part of the real circumference, and if the deviation of the mutual direction of the places from the true meridian be allowed for, within much less.

(13.) Thus we see that with very coarse and rude means of observation and measurement, it is not difficult to arrive at what may be termed a respectable estimate (as contrasted with a mere guess) of the size of our own globe; which is our first step outwards into those distant regions which will next engage our attention. We need not, therefore, dwell on a multitude of intermediate attempts between that epoch and the year 1669, when Picard, under the auspices of the then newly constituted French Academy of Sciences, took up the subject in a truly scientific manner, and with means and appliances of a higher order. They all turned, of course, as every such estimate must do, on the more or less precise measurement of the *length of a degree or a certain number of degrees of latitude* on the earth's surface. But the step which this measure of Picard inaugurated, is distinguished by the abandonment of the old methods of ascertaining such length (viz. by simply measuring it *as an itinerary distance* by rods, or measuring chains, or by rolling wheels self-registering their own revolutions); and substituting for it the infinitely more precise one, which consists in the very careful measurement of a *base line;* the extension from it, northward and southward, of a series of triangles, as above described; the ascertaining, by accurate astronomical observations, the latitudes of the extreme points; and the taking account of the deviation from the true meridian of their mutual direction, by a systematic process of calculation, grounded also on the astronomical determination of their bearings. From that to the present time, this process (in which consists "*geodesy*" as distinct from mere mensuration and surveying) has been generally adopted; with continual improvement of the instruments used; increasing accuracy in all the requisite astronomical observations; and the adoption of a more and more perfect and refined system of computation for the "reduction" of the observations and calculation of the sides of the triangles, *considered as lying not on a plane, but on a spherical surface*, and ultimately (as we shall see) on a spheroidal one. It is not our object to dwell on these details, or to describe more minutely any one of the many operations of the kind which have been carried out or are still in progress in France, England, America, Prussia, Austria, Italy (but more especially and on the vastest scale in the Russian and in our own Indian Empire), and in the southern hemisphere at the Cape of Good Hope. We are only concerned here with the final conclusions arrived at, and with the reasons on which they rest, and these are :—

1st. The length of a degree of the meridian, in whatever region of the earth it is measured, is *very nearly* the same, nowhere varying from a general average by more than about one 200th part of its amount. And from this it follows that the figure of the earth approaches exceedingly near to that of an exact sphere. For the length of such a degree is a measure of the curvature of the surface, it being evident that were any one to travel southward till the meridian altitude of a star was increased by one degree, he must have arrived at a place where the surface on which he stands is just one degree inclined to that at his starting point: so that in walking on he is at that moment pursuing a course deviating by one degree from the direction of his outset. *Now this deviation from a straight course is our idea of curvature.* The curvature of each geographical meridian then is very nearly the same everywhere. In other words, the earth is very nearly a sphere. The average length of a degree of its circumference is 364,578 English feet; or very nearly as many thousand feet as there are days in a year.

2d. Though *nearly*, the degrees are not *precisely* equal. In all geographical *longitudes* the degrees of latitude are found to increase in length in going from the equator towards the poles. An increase in the length of a degree indicates a less amount of curvature. The earth's surface is therefore less curved, or *less convex*—that is, flatter—as we approach its poles on all sides from the equator. Its form then is elliptical, or *oblate*, and its polar diameter somewhat shorter than its equatorial. From the most recent calculations (those of Captain Clarke, founded on a general assemblage of all the measured arcs) it results that the difference of these two diameters is one 292d part of the former.

3d. That the length of its polar diameter is 41,707,796 British imperial standard feet, which is within a single furlong of 500,500,000 such inches.

(14.) Hence it follows, that if we were to increase all our imperial standard measures, each *by one precise thousandth part*,* and designate the measures so increased by the epithet *geometrical* instead of *imperial*, a geometrical inch would be the exact 500,000,000th part; a rod of fifty *such* inches the exact 10,000,000th part of the earth's polar axis; and one of twenty-five *such* (which might be called A GEOMETRICAL CUBIT) the 10,000,000th

* I have before me for ordinary use two *foot rules*, both bought at respectable shops, and not the worse for wear, which differ by more than this amount.

part of its polar *semi-axis:* that is to say, if we disregard so insignificant an error as a furlong upon 8000 miles, or one part in 64,000.

(15.) It follows, moreover (as may be verified by any one who will make the calculation), that if we consent to disregard so trifling an error as one part in 8000; one cubic *geometrical* foot of distilled water at our standard temperature weighs exactly 1000 of our *actual* imperial ounces, and is exactly filled by 100 of our *actual* imperial half-pints.*

(16.) Having thus exhibited the connexion between our ordinary measures of length, weight, and capacity, and the dimensions of the globe we inhabit (a connexion of singular felicity, when we consider the simplicity of the numerical relations), we are prepared to take a further step, and, by using the diameter of the earth itself as a base-line, carry on the same principle of triangulation into our solar and planetary system. In this, the natural unit—that to which astronomers have agreed with one accord to refer all its dimensions—is the mean or average distance of the earth from the sun, or the semi-axis of the ellipse which it describes about that luminary.

(17.) The way in which a knowledge of this distance is obtained being very fully described in our Lectures on "The Sun" and on "Comets," † it is unnecessary to re-

* The deviation of the *actual* French litre and gramme from their true theoretical values, is more than three times as great, being one part in 2730.

† A very unfortunate *erratum* exists in one of the numbers in p.

peat the explanations there given. The use which may be made of Venus as a stepping-stone on the way towards the great centre of our system, however, is there rather alluded to than explained; so that a few words on this subject will not be out of place here. The interval between Venus and the earth when nearest, is not more than one-fourth of the sun's distance, and its angular displacement when seen from opposite extremities of a diameter of our globe, therefore, four times as great as the sun's. Venus, however, is invisible in this situation, except on those very rare occasions (occurring at intervals alternating between eight years and upwards of a century) on which it passes between us and the sun, and is seen as a round black spot on its disc. In this state of things the face of the sun itself serves as a screen on which the planet is seen projected; and its circular outline serves as a celestial line of reference, across which the planet is seen to "transit," as it would across wires fixed in the focus of a telescope; or rather as it would across the circular outline of what astronomers call a "ring micrometer." The sun itself is thus transformed for the time into an astronomical instrument of that description, freed by nature from all the sources of fluctuation and instability which affect our instruments. And the whole observation is reduced to determining the precise moments of time at which the foremost and hindmost borders of the planet cross this

479, col. 1, line 52, of this last-mentioned Lecture as printed originally in "Good Words"—*For* 16,071, *read* 6,071. All the other numbers are right, as they there stand.

ring (which they do in a very leisurely manner), leaving the apparent displacement of the planet *on the sun's disc* to subsequent calculation, on a comparison of reports from all the points of observation selected. One-fourth of the advantage arising from its proximity, it is true, is lost, by the sun itself sharing to that extent in the displacement of the planet; but enough remains to give this a superiority over every other method of measuring the sun's distance.

(18.) Taking as the general conclusion for that distance which we must at present rest in, that assigned in our article last cited, viz., 91,718,000 imperial (or 91,626,282 *geometrical*) miles, we find it equivalent to 23,222 polar semi-diameters of the earth, or ten million times that number of GEOMETRICAL CUBITS of twenty-five *geometrical inches* each.

(19.) Our next step is to the fixed stars, within whose sphere modern science has at length made good a footing, secure, though somewhat unsteady for the present. In conformity with the same principle of procedure, we here rest for our base of operations on our last and greatest *accessible* measured length, viz., the diameter of the earth's annual orbit, a base line of 183,000,000 miles, which, as the orbit is very nearly circular, presents itself (in some situation or other across it) perpendicularly to a line joining the sun and any selected star, so as to be seen unforeshortened from the star. As the earth at half-yearly intervals passes alternately from one to the other extremity of such a diameter, the visual line by which the star is seen will undergo a semi-annual displacement to and fro to the amount of the *apparent* (angular) breadth of the orbit as it would be seen by a spectator in the star. And this, in its equivalent form of annual displacement, is the angle astronomers have to measure for the purpose in question. One would naturally suppose that so enormous a magnitude would be something conspicuous from any distance short of the fabulous; and that here at least we should have something to deal with palpable to very moderate means of observation. Pent up and "chafing within the narrow limit of the world" the astronomer in his measurement of the sun's distance might complain, in the words which the poet puts into the mouth of the great conqueror of antiquity, of restricted elbow-room. Using the world itself as a means of transport, and thus enabled to commence anew on so vast a scale, he might expect to find "ample room and verge enough" for his operations. Quite the contrary! The earth itself seen from the sun would appear as large as the *globe* of Saturn at its medium distance does to us—a very conspicuous object in a moderately good telescope. A globe large enough to fill the *earth's orbit round the sun* would appear to a spectator placed in the nearest fixed star, hardly larger than the third satellite of Jupiter, as seen from the earth; which requires a very good telescope to be perceived to have any *size* at all.

(20.) Two methods only have been devised by which this annual or *parallactic* displacement (as it is technically called by astronomers)* can be ascertained. The first consists in determining the exact situation which its direction in space holds at all times of the year in relation to some plane, and to some line in that plane which we have reason to consider as fixed; or at all events of whose movements (exceedingly small in amount) we can render an exact account. Such a plane is that in which the earth revolves round the sun, or the *ecliptic*, and such a line that of the *equinoxes*, and the astronomical process employed is that by which the two elements technically called the *longitude* and *latitude* of the star are determined. This is in effect the process by which all celestial charts are constructed and catalogues of stars made. Only for this purpose the observations require to be made with the very best instruments; with the minutest attention to everything which can affect their precision; and with the most rigorous application of an innumerable host of "*corrections*," some large, some small, but of which the smallest, neglected or erroneously applied, would be quite sufficient to overlay and conceal from view the minute quantity we are in search of. To give some idea of the delicacies which have to be attended to in this inquiry, it will suffice to mention that the stability not only of the instruments used and the masonry which supports them, but of the very rock itself on which it is founded, is found to be subject to annual fluctuations capable of seriously affecting the result. So that it is only when after a series of observations continued for several years

* What is technically called *parallax*, is only the *half* of the total annual apparent displacement.

it is found that one star appears to be subject to a regularly recurring annual displacement, *such as that which the earth's orbital motion might cause*, and others in its near neighbourhood show no signs of it, we can accept the double conclusion that the one is and the others are not at a measurable distance.

(21.) As the ancients had no knowledge of the earth's motion, so they could have had no conception of this annual displacement of the stars or of their "parallax." Tycho Brahe who rejected the Copernican system, might perhaps have been led to do so from his not having been able to perceive any such displacement of the pole star, ·hich, from the rudeness of his means of observation, he could not possibly have done. Much more lately, in the latter years of the 17th and beginning of the 18th centuries, the attention of many eminent astronomers was drawn, in consequence of the improvements then introduced in the construction of astronomical instruments, to a regularly recurring annual displacement of certain stars observed by them to a very considerable amount, which was at first supposed to be *parallax*, but which proved to be what is now called "aberration," and to be common to all the stars; and when this was recognized finally by Bradley as a result of the motion of light, the idea of a measurable "parallax" was abandoned in despair; to be revived again by Dr Brinkley in 1810; who from his observations with a very fine circle in the Royal Observatory of Dublin *thought* he had detected a parallax of 1″ in the bright star Lyra (corresponding to an annual displacement of 2″). This however proved to be illusory; and it was not till the year 1839 that Mr Henderson, having returned from filling the situation of Astronomer Royal at the Cape of Good Hope, and discussing a series of observations made there with a large "mural circle" of the bright star α Centauri, was enabled to announce as a positive fact the existence of a measurable parallax for that star: a result since fully confirmed with a very trifling correction by the observations of his successor, Sir T. Maclear.

(22.) The parallax thus assigned to α Centauri is so very nearly a whole second in amount (0″·98) that we may speak of it as such. It corresponds to a distance from the sun of 206,265 times that of the sun from the earth, which, as we have already seen, is itself 23,222 polar semi-axes of the latter, thus making a total of 4,789,880,000 such semi-axes (or 10,000,000 times that number of geometrical cubits), equivalent to 18,918,000,000,000 (nearly nineteen billions) of British statute miles. Its near neighbour β of the same constellation and other stars adjacent exhibit no such annual displacement, and are therefore beyond the reach of our measurement. Such, then, is the length of the sounding-line with which we have first touched bottom in the attempt to fathom the great abyss of the sidereal heavens. At such a distance, the vast globe filling the earth's orbit, above spoken of, would be covered from sight by a human hair held at twenty-five feet from the eye.*

(23.) The other mode in which this great question

* Supposing the pupil reduced to a point.

has been approached is to select for inquiry bright stars, which have in their immediate vicinity, so near as to be seen with them at the same time in the same telescope, two or three other very much smaller ones; and without troubling ourselves to determine their *absolute* places in the heavens (so throwing overboard the enormous difficulties which, as we have seen, that determination to a sufficient precision presents), confine ourselves to what may be called a microscopic examination and mapping down of the relative distances and situations of these stars *inter se*. Repeating this at all seasons of the year, we are enabled to ascertain whether the large star maintains steadily the same invariable position among the smaller ones; or is affected by any movements of which they do not partake. There is a general *primâ facie* probability that the brighter stars are nearer than very faint ones: and, near objects being more displaced than distant ones by the spectator's change of place; the large star in the case supposed would appear, by the effect of parallax, to move to and fro among the smaller ones; or rather to describe annually a minute ellipsis among them, the exact counterpart, equal in size and similar in the situation of its longer and shorter diameters, to that into which the earth's orbit itself would be seen projected by the effect of perspective from the star. Now no casual movement, or one arising from any other physical cause, *could* be mistaken for such a motion as this. For, not to mention the completion of the revolution in an exact year, the two diameters of the ellipse ought to stand to each other in a certain definite and calculable proportion known beforehand; and, moreover, the longer ought to be situated in a parallel of latitude, and the shorter in a circle of longitude passing through the star.

(24.) There is a star, the 61st of Flamsteed's list, of those in the Constellation Cygnus—a star far, however, from *conspicuous* for its brightness; being only of the fifth magnitude; but which, for a reason presently to be mentioned, was suspected to be nearer than the generality of the stars. This star was subjected by the late Professor Bessel to the examination above described between the years 1834 and 1838, and the result of his examination (made public by a singular coincidence *a few days before* the announcement of Professor Henderson's discovery) was such as to leave no doubt of the reality of its parallax, to the amount (as slightly corrected by a further continuance of his observations) of 0″·35. Later astronomers,* going over the same ground, with more perfect instruments and improved practice in this very delicate process of observation, have found a somewhat larger result—stated by one at 0″·57, and by another at 0″·51—so that we may take it at 0″·54, corresponding to somewhat less than twice the distance of α Centauri; or to 374,320 solar distances, which light would require about eight years and four months to travel over.

(25.) It cannot be supposed that results like these would be accepted without undergoing the most severe scrutiny and receiving confirmation from further and

* Messrs Auwers and O. Struve.

continued observation. They have received it, and (with exception of those subsequent corrections in the numerical values which we have noticed and included in the above statement) they remain intact, and rank among the well-established facts of astronomy. Moreover, numerous other stars have been subjected to examination, some by one, some by the other method. And the result is not a little surprising. Up to the present time, out of all the stars examined, only a very few exhibit *any* distinctly measurable amount of parallax. The list hitherto accumulated consists only of about ten or at most a dozen. Of these α Centauri, in the southern hemisphere, is the nearest. It is a fine star of the first magnitude, the third or fourth in brightness of all the sidereal host. This is our next neighbour. On the other hand, Sirius, the brightest of all the stars, and Lyra (next to Sirius, one of the four most conspicuous stars in our hemisphere) stand low in the order of proximity. This, of course, only proves that among the stars there exists a very wide range of *absolute* brightness, but by no means invalidates the strong *à priori* reasons for admitting distance as a very important element in determining their relative *apparent* brightness.

(26.) But how, it will be asked, came such a seemingly insignificant object as this No. 61 to be selected for examination at all, to the exclusion or postponement of so many more conspicuous? We reply, by reason of its large apparent *proper motion.* None of the stars we see maintains *quite* the same relative situation among its compeers. It would be strange if it did. Unless nailed to the celestial vault and carried round with it, such a thing is not to be supposed. In the total absence then of any information as to the velocity or direction of the real motions, we can only presume that such as appear to move fastest are nearest. The fact may be otherwise, but such at least is the *primâ facie* presumption. Now, while the stars in general exhibit an annual apparent proper motion averaging less than a second per annum, these two 61 Cygni and α Centauri, are carried annually from their places, by movements apparently rectilineal of 5″·3 and 3″·6 respectively: motions which would carry them away from their places through a space equal to the moon's apparent diameter in 339 and 499 years respectively. In point of fact, we find that they *are* nearer, so that a part at least of their great apparent motions *is* owing to proximity.

(27.) Such a uniformly progressive change of place complicates apparently, but not really, the microscopic process we have described. Being accurately known by long continued observation, both in amount and direction; its effect in displacing the star among its neighbours is easily taken account of and allowed for. The combination of these two motions, the one real and rectilinear, and the other apparent and elliptic, will be readily understood from the accompanying diagram, where *ab*, *bc*, *cd* represent the former continued for three years; *e*, *f*, *g*, the ellipses described in those years in virtue of the latter in the direction of the arrows; and *h i k* the sort of undulating line apparently described in virtue of them both going on together.

(28.) But, besides this, the two stars in question are remarkable in another way. They are both conspicuous DOUBLE STARS, and have been long watched with especial

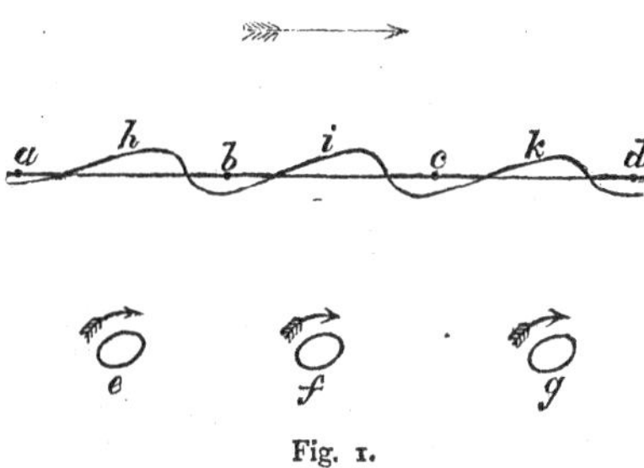

Fig. 1.

interest by reason of the two individuals of which each respectively consists standing to each other in the relation of sun and planet, or planet and satellite. Not only do they keep close company with each other in their journey through space by *a common proper motion;* but while so journeying together they revolve about each other or about their common centre of gravity in regular elliptic orbits, under the influence of that very same law of gravitation which retains the planets in their circulation round the sun. They are, therefore, at the same distance from us, and therefore both equally and similarly apparently displaced by parallax; so that *each of them*, microscopically watched, appears to describe the same minute annual parallactic ellipse above spoken of. This must not be confounded with the elliptic orbit the two stars describe about each other. *That* is on a far larger scale, and requires many years for its completion.

It is to the dimensions of these and similar orbits described by others of those wonderful bodies, the double stars, about each other, that we have now to turn our attention: thus opening another chapter in the history

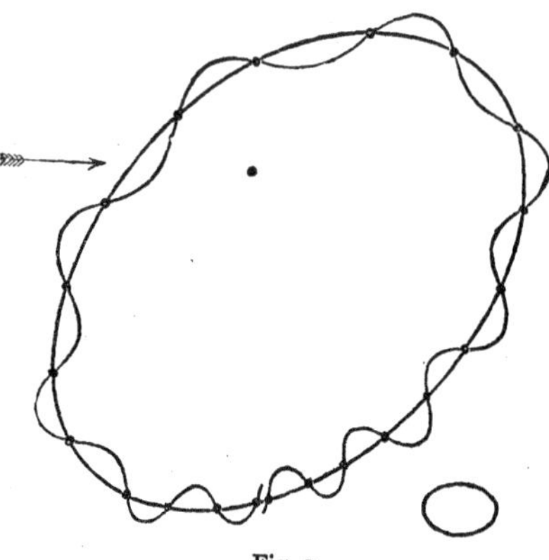

Fig 2.

of sidereal mensuration. The mode in which these two elliptic movements, the larger real, and the smaller apparent or parallactic, are combined together or superposed, and the sort of undulating line apparently described by either star in consequence, will easily be understood by a glance at Fig. 2.

(29.) Assiduous observation, aided by a powerful and not very complicated system of calculation, has enabled astronomers to assign in a great many instances with considerable precision the true *forms* of these orbits as distinguished from those in which, by the effect of perspective (owing to their oblique presentation to our sight), they *appear;* to state the amount of that obliquity;

the situation in space of *the planes* in which they revolve; and the number of years required to complete their revolutions. Among them occurs every variety of form (*always elliptic*), from the nearly circular one of the planetary, to the long ellipsis of the cometary orbits; every variety of oblique presentation, from a plane passing nearly or quite edgeways through the eye of the spectator to one nearly perpendicular to the visual line; and every length of period, from thirty years up to many centuries. The only element about which in the great majority of cases we are left in complete uncertainty, is the actual *size* of the orbit, which cannot become known till the distance of the star is ascertained. For our present purpose then we must confine our attention to those of which at present the distance *is* known. The two just spoken of present a striking contrast. The revolution of the two stars of α Centauri is performed in about seventy-eight years. Their orbit is a very elongated ellipse, decidedly cometary in its character; and its presentation to our sight so nearly edgeways, that the two stars at present almost occult or cover one another; though when at their greatest distance from each other, they would appear, if viewed perpendicularly, nearly thirty seconds apart. The other requires about 514 years for a complete revolution. Its orbit is nearly circular, and its presentation to our view nearly perpendicular: so that we see the distance between the two stars unforeshortened; and so seen it measures almost exactly sixteen seconds, or a little less than the average apparent diameter of the globe of Saturn. Now we have already seen that in the former case the distance between the earth and sun would appear under an angle of $1''$ and in the latter $0''\cdot54$, whence it is easy to conclude that the mean distances of the stars from each other, or the semi-axes of their orbits, are, in the former case about 15, and in the latter about $29\frac{2}{3}$ times that distance. The former orbit would be contained between those of Saturn and Uranus: the latter is about the size of that of Neptune.

(30.) In such orbits, then, gyrating round each other —not in the subordinate relation of sun and planet, but as compeers in dignity and on the equal footing of regal splendour; communicating to each other we know not what benefits, and bound on we know not what errand, —are these wonderful sidereal couples journeying onward through space at the respective rates of 920,000 and 2,500,000 miles *per diem* at the very least: for such would be their proper motions were we sure that *they* are not foreshortened by oblique presentation to our line of sight!

(31.) An interesting, and what to many of our readers will probably appear a very unexpected, conclusion follows from this determination of the distance of these stars, conjoined with the knowledge so obtained of the periodic times of their orbital motion. It enables us to *weigh them;* that is, to state in numbers the proportion which the total *ponderable mass* or *amount* of *gravitating matter* of the two stars of either couple bears to that of the sun, and therefore as a necessary consequence to that of our own globe, and ultimately (if we choose to luxuriate in the long array of figures in which such a calculation would land us) to our British standard ounce, of which this our globe is equivalent to about 210 *quadrillions.**

(32.) It is an elementary proposition in physical astronomy that the time in which two masses so connected into a system by their mutual attraction, revolve about each other in elliptic orbits, depends only on *the sum* of their masses or weights, and on the *length* of the elliptic *relative* orbit, and not at all on its *breadth*, and is therefore the same as if the orbit were circular, *i.e.*, as if the two masses were retained constantly at the same distance from each other, viz., that which we have called their mean distance; and which mean distances, as we have seen in the cases before us, are respectively in round numbers fifteen and thirty times that of the sun from the earth.

(33.) It is an equally elementary conclusion from the theory of gravitation, and was long since demonstrated by Newton, that, so far as the time of revolution is concerned, it is unimportant in what proportions the sum of the masses or the entire ponderable matter of the system is distributed between the two, the distance being unaltered. That time, therefore, would remain unaltered,

* Adopting that nomenclature which calls 1 followed by 6 ciphers a million, by 12 a billion, by 18 a trillion, and by 24 a quadrillion. For the weight of our globe in tons (5852 trillions), see Herschel's "Physical Geography," 2d edit. sect. 5. The elastic forces with which Mr Bailey, in his repetitions of the celebrated "Cavendish Experiment" (from which this estimate of the weight of our globe is concluded) compared that weight, varied from less than one 29,000th part of a grain in some experiments to one 2500th in others! The result, however, being corroborated in various ways, is received without hesitation.

if all the ponderable matter but a single pound were collected in one of them, and that pound circulated about it as the earth does about the sun. Here, then, we have the case stated over again, with only the difference of times and distances, which, in our Lecture on "The Sun," already referred to, § 16, 17, served us to show how we might arrive at a knowledge of the sun's mass, and to calculate that mass. Substitute in the reasoning there explained for one year 78 or 514 years, and for the sun's distance respectively fifteen and thirty times that distance; and the result, in place of the mass of the sun, will furnish us with the total or joint masses of the two stars in the one or the other of these two sidereal combinations or "binary systems" respectively. We shall not trouble our readers with the calculation: suffice it to state the result, viz., that the joint mass in question in the former pair (that in the Centaur), is about $\frac{11}{20}$,—a little more than half that of the sun, or equal to 198,000 earths; and in the latter (in the Swan), about $\frac{1}{10}$ of the sun, equivalent to 36,000 earths.

(34.) Beyond the distances of these two remarkable sidereal combinations, our grasp becomes less and less assured as we push forward into space. Remarkably enough, Sirius and Arcturus, the two brightest stars visible in our hemisphere, stand barely within the limits of any estimation approaching to certainty,—the former being between six and seven, the latter about eight, times the distance of our nearest neighbour in the Centaur. At the distance thus assigned to Sirius, our sun (if any faith can be placed in photometry) would appear as a star hardly of the sixth magnitude—invisible, therefore, or but barely discernible to the ordinary unassisted eye; and it would require four hundred such suns concentred into one to send us the light which that superb star actually does; supposing none lost or extinguished in traversing so enormous a distance: a journey which it would take more than twenty years to accomplish! We speak here only of the proportion between the *lights* of the two bodies; but this can give no indication of that between either their magnitudes or their *weights* or masses, since the *intrinsic* splendour of the surface of the one may, for anything we can tell, exceed that of the other in any proportion. As to the proportion between the masses, however, a very unexpected prospect of being able to ascertain it ere many years shall have elapsed; and even of forming something like a rude estimate of it already, has quite recently opened to us: the history of which may serve to show what persevering industry will accomplish in apparently the most hopeless lines of inquiry.

(35.) Sirius, as the most conspicuous of the stars, has been watched by all astronomers with the utmost assiduity as the principal of the great landmarks of their science; the chief of their list of "fundamental stars;" those to which every observer of necessity resorts to test the stability of his instruments; the rates of his clocks; and every condition which gives precision to his observations. It has long been known, like most and probably all the other stars, not to be absolutely fixed in the heavens: but subject to what we have above described as a "proper motion," or slow progressive movement *proper to itself* as an individual: smaller indeed than those already specified in its apparent amount, but by no means inconsiderable, being sufficient to displace it by about two minutes in angle or one-fifteenth part of the apparent diameter of the moon *per century*: corresponding at the distance of the star to no less an amount of actual linear travel than 1,900,000 miles *per diem*. This movement, in the absence of all apparent reason to the contrary, was of course presumed to be uniform and rectilineal; but as instruments improved, as observations became more exact, and their calculation more scrupulous and refined, this became at first doubtful, and at length demonstrably incorrect. Not to dwell on the steps of the proof, it became apparent that the visible path of the star, mapped down from year to year and from century to century, is not a straight line, but is affected by a small and regularly recurring *undulation*, alternately carrying it to a small distance above and below the medial line, similar to those represented in Fig. 1: the performance of one complete undulation occupying about 49¼ years, and the excursions to and fro on either side of the medial line being about one-sixtieth part of the linear distance passed over in the same interval.

(36.) It was impossible to ascribe *this* phenomenon (as in the case of our star in the Swan) to parallax. Were this its origin, the undulations (as above explained) would be annual, instead of extending over a period of nearly fifty years; and moreover that cause of apparent motion was taken account of in the investigation, in a manner we need not here explain. The only other, and in the then state of knowledge a very obvious, way of accounting for it, was to ascribe these anomalous movements to the attraction of an *unseen* companion; in other words, to consider Sirius as in reality a "double star;" its companion being either a non-luminous body, and in the nature rather of a planet than an associated sun; or, if luminous, so feebly so as to be lost in the rays of Sirius itself, which, in powerful telescopes, is of dazzling brightness. Accordingly, Professor Peters, to whom we owe this interesting investigation, proceeded (by steps which we could not possibly make clear to our readers, and which indeed only *experts* in mathematical calculation can follow) to compute the relative orbit of the pair on the theory of gravitation, and thence to ascertain,—not their *mutual* distance from each other (for that necessarily *then* remained uncertain) but that of Sirius itself *from their common centre of gravity*. For this he found an apparent angular measure, of $2''\cdot4$, corresponding to about $16\frac{1}{4}$ times the distance of the earth from the sun; and calculating on his final result, the observed anomalous deviations from uniform rectilinear motion were found to be satisfactorily accounted for.

(37.) It is now time, however, to mention what, to render our explanation more simple, we have hitherto kept out of view, viz., that all the foregoing calculations were directed only to that part of the "proper motion" of Sirius which carries it in the direction of a parallel to the earth's equator, or, as it is technically called, "in right ascension." If that movement were coincident in direction with such a parallel, there would remain nothing further to explain. But such is not the case. It is oblique; and may therefore be regarded as composed of two movements,—the one along that parallel (in right ascension), the other perpendicular to it, or, as it is technically called, "in declination." Now these movements admit of a distinct and separate examination, and it is clear that, if *both* do not agree in indicating the same kind of undulation and the *same identical period*, the explanation so afforded of what may be called one half of the phænomenon is at variance with that of the other. Mr Peters left this other half untouched; but very recently that also has been examined by an American computist, Mr Safford, on the same principles; and the result is that the orbital motion, which accounts for the one set of movements, gives at the same time a sufficiently satisfactory explanation of the other.

(38.) Here, then, we are furnished with another example like that afforded by the grand discovery of the planet Neptune by the calculations of Adams and Leverrier. The existence of a celestial body not seen and not before known to exist, has been revealed to us and its orbit computed, by the simple application of mathematical calculation grounded upon observed irregularities in the movements of one already well known.

(39.) The parallel of the cases promises to be still closer. Neptune, as is well known, was immediately sought and found in the place assigned to it by the calculation. In January 1862, Mr Alvan Clark, an eminent optician of New York, turning on Sirius a fine telescope of his own construction, noticed extremely near to it a minute star which had eluded all former observation. This *may be* the body in question. There is even some reason to suppose *it is.* Its apparent situation is stated to be at least not such as to be incompatible with such a connexion. Its *real existence has been verified*, and its apparent distance from Sirius measured, and found to be about seven seconds ; corresponding (if seen unforeshortened) to about forty-seven times the distance of the sun from the earth.

(40.) Another beautiful specimen of these binary sidereal systems is presented by the star No. 70 in Flamsteed's list of those in the constellation Ophiuchus, and therefore cited as 70 Ophiuchi. The ellipse described by the stars of this pair (the one a star of the fourth, the other of the sixth, magnitude) has been determined with much care and every probability of considerable precision. The period of their mutual circulation may be stated at about ninety-six years, and the semiaxis of their mutual ellipse *in angular measure* at 4″·8. Of this elegant couple the parallax has been ascertained by M. Krüger, from observations made in 1858 and 1859, at 0″·16. And from these data he concludes in the very same way:—First, their distance from our own system (1,272,000 semi-diameters of the earth's orbit) ; secondly, the mean distance of the stars from each other ($30\frac{1}{2}$ such semi-diameters, so that here also their relative orbit is nearly equal to that of Neptune) ; and, thirdly, the total *mass* (equivalent to $3\frac{1}{10}$ times that of the sun).

(41.) Shall we spread our wings for a farther flight—to the region of the nebulæ? For such an excursion we are hardly yet prepared. Our present reach extends, as we have seen, only to a very few of our nearest neighbours among the *stars*; a class of bodies which we have every reason to believe form with our own sun a system—to us a universe—but which, removed to the distance of the nebulæ, would appear perhaps as one of them. Moreover, it is not wings, but a resting-place for the sole of our foot that we want. If we knew in what orbit the sun itself is moving (for that it moves is certain, and with no trifling speed), and if human observations were to endure till it had completed half a circuit in that vast orbit—then indeed we should have established a new base line from whose extremities the *parallax* of the nearest nebula might become sensible. Failing this, we must rest content with such probable indications as we can glean from other sources.

(42.) There is one which can hardly fail to strike any one who does not reject altogether from his philosophy the consideration of design and purpose in the construction of the frame of nature. In their orbits round the sun, the earth and other planets carry round with them satellites retained in *their* orbits by gravitation to their primaries. These orbits, though very sensibly disturbed by the sun's attraction, are yet in no case so much so as to hazard in the smallest degree the stability of these miniature planetary systems, or in the lapse of even indefinite ages to produce any very material change in their relations to their primaries or to each other. The enormous distance which separates the sun from the nearest fixed star affords a still more complete guarantee against the possibility of any disturbance of the planetary movements by *their* attraction, and may not unnaturally be considered as so intended. A continuance of the same system of precaution (if we may venture on the use of such a word) against external influence, into the mutual relations of *sidereal* systems might therefore lead us to expect that the intervals between them would at least bear some very large proportion to the extent of each. That there exist instances of nebulæ which appear to be bound together by a kind of companionship similar to that of the double stars, does not in the least invalidate this as a general conclusion. Here, however, figures avail us nothing. Nor can it be necessary, after what has been already said, to stimulate our imaginations to any further effort to grasp and comprehend distances and magnitudes inconceivable by man. Suffice it that in the dim glimpse thus caught of an immensity of material existence stretching outward by steps continually more and more gigantic, we carry with us not a mere general impression, but a well-founded conviction grounded on an induction from observed facts of measurement and computation, that the same mechanical laws at least; the same relation between matter, force, and motion as those we see in action around us, prevail in the uttermost regions of space; and regulate, there as here, the evolutions of the systems disseminated through it. In the endless variety of combination exhibited among the double stars too (to say nothing of a multitude of other phænomena, *less clearly intelligible*, which the sidereal heavens present, and to which our subject has not led us to refer), we trace the same inexhaustible fecundity of design realized and embodied in the same unity of workmanship which in this our planetary system we find luxuriating in so surprising a variety of forms, magnitudes, and mutual relations among its primaries, satellites, rings, comets, and asteroids.

(43.) Is the material universe finite or infinite? The question is as old as Aristotle; and the answer, though *unanswerable*, never yet convinced mortal man. A material universe must consist of material objects, each individual of which, being a really existing thing, must possess that attribute of all real existing things, place. Every two objects then, be they where they will at any certain moment of time, mark *two definite places*, and the distance between them, or the straight line joining them, has two definite *terminations*. It is not therefore infinite in length, but finite, *i.e.*, terminated. Now an assemblage of objects, *every two* of which are distant from each other by a finite interval, cannot be infinite in extent. The speculation is unprofitable enough in itself, and the difficulty it involves turns on the mental substitution of a positive and conceivable notion of "the infinite" for the purely negative and utterly inconceivable one which it carries with it into all matters where the term is employed in its logical sense. Our only reason for at all alluding to it is, that to *us*, practically speaking, the material universe must be regarded as infinite: seeing that we can perceive no reason which can place any bounds to the further extension of that principle of systematic subordination which we have already traced to a certain extent; and which combines in its fullest conception a unity of plan and singleness of result with an unlimited multiplicity of subordinated individuals, groups, systems, and families of systems. Thus it by no means follows that all those objects which stand classed under the general designation of " nebulæ " or " clusters of stars," and of which the number already known amounts to upwards of five thousand, are objects (looked upon from this point of view) *of the same order.* Among those dim and mysterious existences, which only a practised eye, aided by a powerful telescope, can pronounce to be *something different from* minute stars, may, for anything we can prove to the contrary, be included *systems of a higher order* than that which comprehends all *our* nebulæ (properly such) reduced by immensity of distance to the very last limit of visibility. And this conception, we may remark, affords something like a reasonable answer to those who have assumed an *imperfect transparency* of the celestial spaces, on the ground that, but for some such cause, the whole celestial vault ought to blaze with solar splendour, seeing that in no direction of the visual ray, if continued far enough, would it fail to meet with a star. Such would no doubt be the case were all space occupied by stars disseminated through it uniformly, *i.e.*, so that the same number of stars should in every region be comprised in the same space. But no such consequence would follow were the law of sidereal distribution such as we have been here describing: a position, however, with whose demonstration (founded on the very elementary properties of a decreasing geometrical progression) we shall not trouble our readers.

(44.) Such a speculation as this just mentioned may possibly appear irrelevant. But it must be remembered that it is LIGHT, *and the free communication of it from the remotest regions of the universe*, which alone can give, and does fully give us, the assurance of a uniform and all-pervading energy—a mechanism almost beyond conception complex, minute, and powerful, by which that influence, or rather that movement, is propagated. Our evidence of the existence of gravitation fails us beyond the region of the double stars, or leaves us at best only a presumption amounting to moral conviction in its favour. But the argument for a unity of design and action afforded by light stands unweakened by distance, and is co-extensive with the universe itself.

LECTURE VI.

ON LIGHT.

PART I.—REFLEXION—REFRACTION—DISPERSION—COLOUR—ABSORPTION.

N a conversation held some years ago by the author of these pages with his lamented friend, Dr Hawtrey, Head-Master and late Provost of Eton College, on the subject of Etymology, I happened to remark that the syllable *Ur* or *Or* must have had some very remote origin, having found its way into many languages, conveying the sense of something absolute, solemn, definite, fundamental, or of unknown antiquity, as in the German words *Ur-alt* (primeval), *Ur-satz* (a fundamental proposition), *Ur-theil* (a solemn judgment)—in the Latin *Oriri* (to arise), *Origo* (the origin), *Aurora* (the dawn)—in the Greek Ὅρος (a boundary, a mountain, the extreme limit of our vision, whence our *horizon*), Ὁράω (to see), Ὀρθὸς (straight, just, right), Ὅρκος (an oath or solemn sanction), Ὧραι (the seasons, the great natural divisions of time), &c. "You are right," was his reply, "it is the oldest of all words; the first word ever recorded to have been pronounced. It is the Hebrew for LIGHT (אור AOR)."

(2.) Assuredly there is something in the phænomena of Light; in its universality; in the high office it performs in creation; in the very hypotheses which have been advanced as to its nature; which powerfully suggests the idea of the *fundamental*, the *primeval*, the antecedent and superior in point of rank and conception to all other products or results of creative power in the physical world. "It is LIGHT," as we took occasion to observe at the conclusion of the last lecture (not without reference to this very consideration), "and the free communication of it from the remotest regions of the universe, which alone can give, and does fully give us, the assurance of a uniform and all-pervading energy—a MECHANISM almost beyond conception complex, minute, and powerful, by which that influence, or rather that movement, is propagated. Our evidence of the existence of gravitation fails us beyond the region of the double stars, or leaves us at best only a presumption amounting to moral conviction in its favour. But the argument for a unity of design and action afforded by light stands unweakened by distance, and is co-extensive with the universe itself." *

(3.) What we propose in the following lecture is to make intelligible, in as simple language and form as the nature of the subject will admit, the grounds of this assertion. In some of its features it is too complex and abstruse to be thoroughly followed out by any one

* "Celestial Measurings and Weighings," p. 218.

not familiar with some of the most intricate departments of mathematical science. In explaining such features (when unavoidable), without prejudice to the strictness of mathematical reasoning adducible and held to be conclusive and satisfactory by those who have mastered it, we must have recourse to analogies more or less close with processes we see going on in nature; and which, whether perfectly understood or not in their *modus operandi*, we, at all events, perceive to consist in a sequence of events, comprehensible in themselves and arising naturally and familiarly one out of another. There are many phænomena of polarized light which admit of being so, as it were, shadowed forth to the mind of a beginner as analogous to things familiar enough. In such cases, though the analogy may be imperfect, or even altogether incompetent to stand for an explanation, the phænomenon is sometimes so neatly conveyed to the intellect, that by generalizing to the extreme all the terms used in describing the one, it is very conceivable that the cardinal feature of the other—that which dominates its whole explanation—*may* be included. Even if not so, the object is so far answered, that the student remains possessed of a mental picture which will not allow him to forget its prototype. And it is not a compendium of Optics, or an essay on Vision, or an account of telescopes, microscopes, or other optical instruments, that he has here to expect. Nothing of the kind could by possibility be comprised within such limits as a contributor to a work of this kind must necessarily observe. Suffice it to convey to his apprehension some idea of at least the general nature of the mechanism by which it seems now agreed, with hardly a dissentient voice, that the peculiar communication between distant objects which we call light is effected; and by which, or by *some* mechanism of a nature still more recondite, and at present perhaps beyond our conception of possibility, it must be so.

(4.) That we see, is proof of a communication of some sort between the eye and the thing seen. That we cannot see in the dark, is proof that such communication is not the mere act of the eye. And that one object is capable of impressing a photographic picture of itself on another, is proof that the eye, though essential to *seeing*, has nothing whatever to do with the process by which such communication is performed. And furthermore, the immense variety and extent of the chemical agencies of light as displayed in its action both on organic and on inorganic matter, revealed to us by the late discoveries in photography, assign to it a rank among natural agents of the highest and most universal character; and have even rendered it exceedingly probable, if they have not actually demonstrated, that vision itself is nothing but the mental perception of a chemical change wrought by its action on the material tissue of the retina of the eye.

(5.) At all events, it is not by any sympathy, or *absolute direct relation* between the eye and the object, that the latter is seen. The intermediate space, and indeed *all space*, is concerned in the process. An object is not seen unless it be in a certain state, which we call "luminous:" a state either natural to it, as in the flame of a candle or the sun; or induced, by being placed in presence of another luminous object, as when a sheet of white paper is laid in the sun or before a candle. Nor is it then seen if a screen of metal or any of the class of substances called "opake" be interposed anywhere in the direct straight line of communication; while on the other hand, when so hidden from direct vision, it may be rendered visible "by reflexion" from a polished surface held at a fitting angle, *anywhere out of that direct line*, provided only such surface be not similarly screened either from the object or from the eye. Thus we learn two things:—First, that the line of *uninterrupted* luminous communication is a *straight one;* and, secondly, that any point whatever in a sphere of indefinite radius surrounding a luminous object (in other words, in infinite space) may become included in the line of indirect or deflected luminous communication between any two places. *The agency*, whatever its nature, *is there and ready*, requiring only a fitting arrangement of material and tangible substances to make it available.

(6.) Light, though the cause of vision, is itself invisible. A sunbeam, indeed, is said to be seen when it traverses a dark room through a hole in the shutter—or when in a partially clouded sky luminous bands or rays are observed as if darted through openings in the clouds, diverging from the place (unseen) of the sun as the vanishing point of their parallel lines seen in perspective. But the *thing seen* in such cases is not *the light*, but the innumerable particles of floating dust or smoky vapour which catch and reflect a small portion of it, as when in a thick fog the bull's-eye of a lanthorn seems to throw out a broad diverging luminous cone, consisting in reality of the whole illuminated portion of the fog. The moon is seen in virtue of the sun's light thrown upon it. Where the moon *is not* we see nothing, though we are very sure that when in the course of its revolution it shall arrive in the place we are looking at, we shall see it, and that if our eyes could be transferred to the moon's place, wherever it may be in the firmament (if not eclipsed), we should *from it* see the sun. There then, at all times, *is* the light of the sun, but not *visible as a thing.* It exists as an agency. What is true of the sun is no doubt equally so of a star; so that when we look out on a dark night, though we are sure that all space is continually being crossed in every direction by the lines of its communication, *along all which it is active;* and in particular, that all the dark space immediately around us (outside of the earth's shadow) is, so to speak, flooded with the sun's light, we yet perceive only darkness, except where our line of vision encounters a star.

(7.) What then *is* LIGHT? or, in other words, what is the nature of that communication by which not only information is conveyed to our intellectual and perceptive being; but chemical and various other changes are operated even on inorganic matter by processes originating as it would seem in sources situate in the most distant regions of space (for, be it observed, it has been clearly proved that the light of the stars does produce photographic effects powerful enough to imprint their images permanently on surfaces duly prepared to receive them). Is there any physical mode of conveyance by which, reasoning from what we see in cases which we are able to analyze, we can imagine either a material agent to be bodily transported, or a movement propagated, or an influence wafted, from place to place, so as to render a rational and consistent account of the phænomena of light, or so at least as, generalized, and (so to speak) sublimated in modes not inconsistent with the known properties of matter, to do so?

(8.) One feature is common to all ordinary physical modes of communication. The transmission from place to place, be it of what it may—of a letter by post, a gunshot, a sound, a wave, a tremor, or a shock even of an earthquake—*occupies time.* It *has a velocity:* sometimes a very great, but anyhow a measurable one. Is this the case with light? The answer, from all ordinary experience, would be in the negative. But this is only because the velocity in question is so great that the longest distances to which we can send a flash of light and receive it back again by reflexion is traversed in an interval of time too short to be perceived *as* an interval, so that the reflexion *appears* to be simultaneous with the direct flash. It is otherwise when we bring to bear on the question the ingenious combinations and delicate appliances of modern science. The telescope enables us to become eye-witnesses in the way of astronomical observation of events which take place at distances in space almost inconceivably greater than any we can measure here on earth; at *times* calculable beforehand.

And in the way of experiment, the contrivances of clock-work enable us to register the subdivisions of what we call "an instant" into hundreds, nay, thousands, of equal and exactly measurable portions—applying, so to speak, a microscope to time, and estimating, by undeniable calculation, portions of it utterly eluding all our powers of perception. The question has been asked in both these modes, by astronomical observation and by direct physical experiment, and the answer, from each, has been affirmative; and from both agreeing, in a manner which may well be considered wonderful.

(9.) The planet Jupiter is attended by four satellites which revolve round it in orbits very nearly circular, and whose dimensions, forms, and situations with respect to that of the planet itself are now perfectly well known. The periodical times of their respective revolutions are also ascertained with extreme precision, and all the particulars of their motions have been investigated with extraordinary care and perseverance. The three interior of them are so near the planet and the planes of their orbits so little inclined to that in which it revolves round the sun, that they pass through its shadow, and therefore undergo eclipse, at every revolution. These eclipses have been assiduously observed ever since the discovery of the satellites, and their times of occurrence registered. As they afford a means of determining the longitudes of places, the *prediction* beforehand of the exact times of their occurrence becomes an object of great importance: and it is evident enough that, all the particulars of their motions being known (as well as of that of the planet itself, and therefore of the size and situation of its shadow), there would be no difficulty in making such prediction (starting from the time of some one observed eclipse of each as an epoch); *provided always* each eclipse were *seen at the identical moment when it actually happened.* Moreover, on that supposition, the times *recorded* of all the subsequent eclipses ought to agree with the times so *predicted.* This, however, proved not to be the case. The observed times were sometimes earlier, sometimes later than the predicted; not, however, capriciously, but according to a regular law of increase and decrease in the amount of discordance, the difference either way increasing to a maximum,—then diminishing, vanishing, and passing over to a maximum the other way, and the total amount of fluctuation to and fro being about $16^m\ 27^s$. Soon after this discrepancy between the predicted and observed times of eclipse was noticed, it was suggested that such a disagreement would necessarily arise if the transmission of light were not instantaneous. This suggestion was converted into a certainty by Roemer, a Danish astronomer, who ascertained that they always happened earlier than their calculated time when the earth in the course of its annual revolution approached nearest to Jupiter, and later when receding farthest: so that in effect the extreme difference of the errors or total extent of fluctuation—the $16^m\ 27^s$ in question—is no other than the time taken by light to travel over the diameter of the earth's orbit, that being the extreme difference of the distances of the two planets at different points of their respective revolutions. At present, in our almanacs a due allowance of time for the transmission of light at this rate, assuming a uniform velocity, is made in the calculation of these eclipses; and the discrepancy in question between the observed and predicted times has ceased to exist.

(10.) Taking the diameter of the earth's orbit, as concluded from the sun's observed parallax,* at 24,000 diameters of the earth itself, and the latter diameter at 7925¾ miles,† this gives a velocity of 192,700 miles per second.

(11.) So vast a speed seemed at first incredible; to some indeed even more so than an instantaneous communication. The one *might* be conceived as the result of some sort of spiritual communication: the other seemed, in those days, to transcend all imaginable limits of mere physical agency. But it soon received a very unexpected confirmation from Dr Bradley's discovery of the ABERRATION of light: to conceive which, let any one imagine a long tube held perpendicularly, at perfect rest, while a falling body (a drop, suppose of a shower of rain), descending also perpendicularly, should pass down its axis. If it entered at the centre of its upper orifice, it would issue at that of the lower; and, judging from this indication alone, and knowing the tube to be exactly vertical, a spectator would truly conclude from it that the descent of the drop was so also. Supposing him and

* See p. 196, note.

† This is the equatorial diameter.

the tube, however, to be carried along uniformly in any direction by a movement unperceived by himself, the hinder part of it would advance to meet the falling drop; which would, if the movement in advance were sufficiently rapid, cause it to strike against it; or if not, to emerge at the lower end so far behind the centre as that movement had carried the tube during the time of its passage from end to end. And this deviation would obviously bear the same proportion to the length of the tube that the velocity of the falling drop bore to that of the tube's advance. Judging, then, from this indication alone, if unaware of his own motion; he would conclude the fall of the drop to be inclined backward from the perpendicular by a certain angle—but if, suspecting it, he should reverse his movement, and travel with equal speed the contrary way, he would find an equal deviation in the contrary direction, and would thus arrive at the *certainty* that it was to the motion of himself and the tube, and not to any real obliquity in the fall of the drop, that this apparent deviation was owing. And by measuring its angular amount (which would be easy by the help of the marks left by two drops in the opposite circumstances on a screen at the lower end of the tube, and comparing it with the length of the latter), this angle, which might be called the *Aberration* (from perpendicularity) of the *apparent* line of fall, would inform him of the proportion his own velocity bore to that of the drop in its passage, and, the former being known, would enable him to estimate the latter.

(12.) All this is a paraphrase of the astronomical phænomenon in question. The rain-drop is the light; the tube, a telescope; the screen at its lower end, a micrometer; and the two opposite directions of the observer's motion, the two tangents at opposite sides of the earth's orbit at right angles to the situation of a star as viewed from either. And the angle in question is what astronomers call their "Constant of Aberration"—a very minute one indeed, but perfectly well measurable—amounting to about a third of a minute (20″·45), from which it results that the velocity of light is about ten thousand (more exactly 10,089) times that of the earth in its orbit, which we know to be very nearly 19 miles (18·923) per second, which gives 190,860 miles per second for the velocity of light.

(13.) Two different experimental processes for measureing this velocity have been devised and executed—the one by M. Fizeau, of the Parisian Academy of Sciences; the other by M. Léon Foucault, recently and most deservedly elected into the same illustrious body; the inventor of that elegant instrument, the Gyroscope. Both depend on the principle that the impression left on the eye by any luminous object persists for a sensible, though very minute, time (about the tenth of a second); so that an object presented to the sight by successive glimpses only, following each other more frequently than ten times in a second, is seen continuously. If only just so frequent, a fluttering is perceived; but this diminishes as the rapidity of presentation is increased: and when much more frequent, distinct and perfectly uninterrupted vision is produced. In M. Fizeau's experiment (which is the simplest in its conception and explanation), these glimpses are obtained by looking through an opening in a screen corresponding exactly in size and shape to one of the intervals between the teeth of a metallic wheel which is made to revolve before the opening, so that as the teeth pass in succession, they intercept the light so long as they cover it; but allow it to pass when, in place of a tooth, an interval is presented before the eye. Imagine such a wheel, screen, and opening, the wheel being at rest in the last-described situation; and through another such an opening in the same screen, corresponding exactly in size, shape, and situation to another of the intervals between the teeth, let a sunbeam be directed outwards, in a direction parallel to the axis of the wheel, by a highly-polished reflector, so as to strike upon another such reflector so placed at some considerable *and measured* distance from the wheel, that the light shall be reflected back again by this second mirror. By slightly inclining and properly adjusting this it may be made to return, not to the orifice from which it issued, but to the other behind which the eye of the observer is placed. In this state of things, when all is at rest, he will see the reflected light; but if the wheel be turned slowly round, a tooth will come before the first reflector in place of an opening, and intercept the light—then another opening, another tooth, and so on, producing successive glimpses of light separated by dark intervals.

(14.) If the motion of the wheel be gradually accelerated, so that more than ten teeth pass before the orifice in a second of time, these glimpses run together into continuous vision; and if considerably more numerous (suppose fifty or sixty per second), the light is perceived steadily as if the wheel were at perfect rest—only, however (if the intervals between the teeth be exactly equal to the breadths of the latter), of half the brilliancy, seeing that only half the quantity of light will have entered the eye in the same time. The motion of the wheel still continuing to be accelerated, however, when it has attained a certain very great rapidity the light is gradually perceived to grow feebler and at length altogether disappears. This happens when the velocity of rotation is such as to bring a tooth of the wheel precisely to cover the whole of the orifice in the screen into which the returning beam should be delivered at the very moment of its arrival, so closing it up altogether; that is to say, when the rotation is just so rapid as to carry each tooth over its own breadth during the time taken by the light to go and return. When this happens, suppose the *acceleration* of the wheel to cease, and its motion to be maintained uniform. Then by counting the turns made per minute by the driving-handle of the train of wheel-work, or otherwise registering its speed; and knowing (from the construction of the train) how many turns of the wheel correspond to one of the driver, as also how many teeth it carries, the exact duration of this interval, no matter how minute, can be exactly computed, so that the time and the space run over by the light in that time both become known.

(15.) If the rotation be now still further accelerated,

the light begins to reappear, and gradually increases to its former brightness, in which state of things the obstructing *tooth* has been carried, in that same interval of time, quite clear of the opening, and the next *notch* brought exactly opposite to it. With yet increased speed, the light again vanishes, again reappears, and so on alternately, as the second, third, or fourth tooth or notch is successively brought before the opening; and on comparing the velocities of rotation corresponding, they are found to increase in arithmetical progression; which obviously ought to be the case. In M. Fizeau's experiments,* the distance between the reflector and the revolving wheel was about 8600 metres, thus giving for the whole distance travelled over by the light going and returning 17,200 metres, or about $10\frac{3}{4}$ miles; and for the time occupied in its journey, hardly more than the 18,000th part of a second. A velocity of 196,000 miles per second was assigned by him as their final result, exceeding by about one-sixtieth part that resulting from the astronomical observations.

(16.) The experiments of M. Foucault, however, leave no doubt that this last result is too great. In these experiments, instead of measuring these minute intervals of time by the rotation of a toothed wheel, a revolving reflector was employed in pursuance of an idea suggested by Mr Wheatstone, and applied by him

* The actual details of this experiment, as executed by M. Fizeau, were somewhat more complicated. Telescopes were used, &c. For clearness of explanation, we have reduced the whole process to its simplest form of expression.

to measure the velocity of electricity. Without figures, and without much more verbal detail than would be compatible with our limits, it would be impossible to give a clear conception of the conduct of this delicate and refined experiment. Suffice it to state, as its ultimate result, a velocity of 185,172 miles per second.* As there are other and independent reasons for believing that the sun's distance has been over-rated by about one-thirtieth in our estimate of 12,000 diameters of the earth, and that, in consequence, the velocity of light deduced from the phænomenon of aberration ought to be diminished in the same proportion (which would reduce it to 186,300 miles per second), we are authorized to conclude that in estimating this velocity at 186,000 miles we are within a thousand miles of the truth.

(17.) The form of experiment proposed and executed by M. Foucault has this great advantage over the other, —that it can be carried out within much smaller limits of distance. A few yards of travel suffices for the determination of this enormous speed. And this makes it possible to compare the velocity of light in its passage through air and water, and other transparent liquids—*with this remarkable result, that the rate is found to be slower in the denser medium ;* a result of the utmost importance, as we shall presently see, as a *crucial fact* in deciding between the claims of the two great rival theories of light to be received as valid.

* 298 millions of metres. See *Comptes Rendus de l'Institut*, Sept. 22, 1860.

in like manner have to be reversed on interchanging the illuminating and illuminated points. On neither supposition could the same *intrinsic* law of communication carry the ray from A through P, to B, and from B, through P, to A. This, then, is the law of regular reflexion, commonly expressed by saying that *the angle of incidence is equal to that of reflexion and lies in the same plane with it.*

(20.) If the reflecting surface be a plane, there will be only one point in it which fulfils these conditions. Thus a perfectly polished flat surface of silver, free from scratches, or that of still water, sends no light to the eye from a candle, and is in fact invisible, except at this one point so determined whence the light is reflected to the eye, and in the direction of which from the eye the reflected candle is seen. With curved surfaces, as well as with those we designate as "rough" or "unpolished," the case is different. In all surfaces of this last-mentioned description the microscope reveals to us such irregularities, such innumerable and abruptly broken facets, protuberances and hollows, as to satisfy us that in every, the most minute, visible portion of such a surface, places must occur in which the condition of equal inclination of the two lines in question to the actual surface, *as it exists in those places,* is satisfied—so that a ray there reflected may reach an eye however situated. By such rays, and by others which have entered into the substance of the object and been there internally reflected or otherwise bent, in a manner presently to be explained, all surfaces not self-luminous become *visible* as *objects,* being seen by rays "scattered" from them in every possible direction.

(21.) It is to this power of "scattering" the incident light in all directions, then, that surfaces owe their visibility, and that by its aid we are enabled to trace the course of a ray of light itself *as if* it were a visible thing. Thus a sunbeam passing through a small hole and received on smoke is *seen*, and on a white screen moved rapidly to and fro behind it, appears as a straight luminous line or beam, by the momentary persistence of the sensation caused in the eye at every successive point of its motion; and so, after reflexion or refraction, may its subsequent course be rendered matter of ocular inspection. A pleasing and elegant experiment is to hold a common reading-glass (or even a spectacle-glass) in the sun, and to move rapidly to and fro behind it a white paper, when the course of the refracted light, *converging* from all parts of the glass to the "focus," will be seen in the air as a solid luminous cone, having the glass for its base and the focus for its apex.

(22.) The reflexion of light, whether "regular" or "scattered," is, except under very peculiar circumstances to be presently noticed, only partial; so that the reflected image of an object is seen fainter and less luminous than the object itself directly viewed. This is perceptible in an ordinary looking-glass; yet more so when the reflecting surface is still water, or unsilvered glass. The *most reflective* substances are the white metals—such as silver, speculum-metal, steel, or quicksilver: transparent or semi-transparent bodies being much inferior in respect of this quality. If the substance on which the light falls be of the kind called opake, the reflected is the only portion which can be rendered sensible to sight or otherwise traced. But if transparent, a very remarkable phænomenon occurs. The incident ray is, as it were, split or subdivided at the point where it meets the surface of the body; one portion pursuing its subsequent course outside of it, as a reflected ray, in the manner above described; the other within it, undergoing what is called "*refraction*," being bent aside from its former direction at its point of entry, after which it pursues a straight course within the substance or "medium."

(23.) If the "refracting medium" be a liquid, a glass, a jelly, or any substance in which no indications of inequality of internal texture can be discovered—no signs of lamination or "grain" shown by a greater tendency to split or "cleave" in one direction more than another, this intromitted portion is *single*. The whole of the refracted light pursues its course from the point of its entry as one ray. The same is also the case when the refracting medium belongs to the class of bodies called "crystallized," or which present a definite "cleavage;" provided the "primitive form" of their crystals be either a cube, a regular octohedron, or a rhomboidal dodecahedron, such as rock-salt, alum, or garnet. In all other transparent crystals the intromitted portion of the light divides itself from the moment of its entry into two distinct rays, pursuing different courses, and presenting the phænomenon known under the name of "double refraction," such substances being called "doubly-refractive media," of which the substance called Iceland Spar, or crystallized carbonate of lime, offers a beautiful example. And here we may pause for a moment to observe that at this point we already find ourselves introduced to an assemblage of relations between light and material objects, which divide the whole universe of such objects, infinite as they are in variety, into *classes*, characterized by their habitudes with respect to light in its reflexion from and passage through them. The importance of this remark will grow upon us as we advance further into the subject, and come to perceive that the classification of bodies according to their "optical properties" stands in direct connexion with their most intimate peculiarities of mechanical structure and chemical constitution; and brings us, so to speak, into contact with all those more recondite properties and reactions of the ultimate particles of bodies which constitute the domain of molecular physics.

(24.) Confining ourselves now to the case where the refraction is single, the rule which determines the course of the refracted ray is as follows. Suppose at the "point of incidence" (*i.e.*, where the ray first enters the medium) a line be drawn perpendicular to the surface. Then, first, the refracted ray will lie in the same plane which contains both the incident ray and this perpendicular; and, secondly, the ray will be so bent at that point that the exterior and interior portions shall make with the perpendicular, not equal angles as would be the case were there no flexure, but angles so related that their *sines* (not the angles) shall bear to each other a certain invariable proportion, whatever be the angles themselves, or whatever be the obliquity of the incident ray to the surface. As it is quite essential to the understanding of what follows that this, "the law of ordinary refraction," should be clearly apprehended, we will illustrate it by a figure. Let A C B be a section of the surface

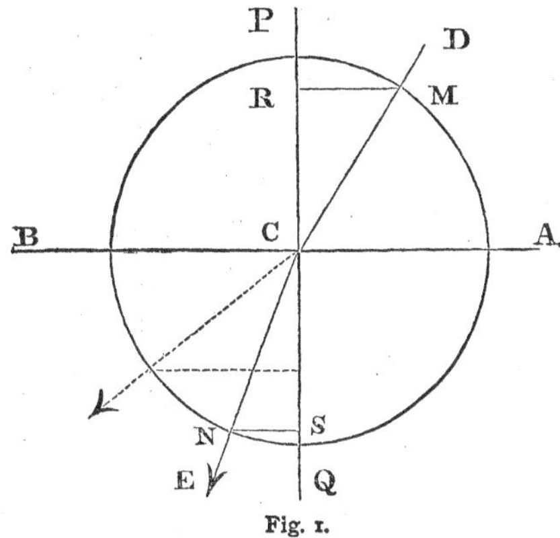

Fig. 1.

by the plane in which the ray D C, incident at C, and P C Q the line perpendicular to the surface at C, both lie, and C E the refracted ray. Taking C for a centre, with any radius, C M, describe a circle cutting the incident and refracted rays in M and N, from which points draw M R, N S perpendicular to P C Q. Then will these two lines be to each other, in one and the same invariable proportion, whatever be the inclination of the original ray D C to the surface, or to the perpendicular C P. This latter inclination is what is understood by the "*angle of incidence*," and the corresponding inclination (to the perpendicular Q C P) of the refracted ray, by "*the angle of refraction.*"

(25.) It is evident from what we said in the last paragraph, that according to the greater or less disproportion between the lines M R, N S, on the diagram there given, or the sines of the two angles of incidence and refraction, the greater or less will be the amount of bending (or *angle of deviation*, as it is called) of the ray at its point of transmission, for one and the same degree of obliquity—as also that for one and the same medium, the *deviation* increases *with* the angle of incidence (though not *proportionally* to it) being *nil* when the ray enters perpendicularly, and a maximum when just grazing the surface. If in any case M R be greater than N S, or the "ratio of the sines" be one of "greater inequality," the bending will be *towards* the perpendicular; if less, or if that ratio be one "of less inequality," *from* it; as indicated by the course of the dotted ray in the figure. If the former be the case in any instance, as in that where a ray passes out of air into water, the latter will happen in the reverse case, as where it passes out of water into air: that is to say, in optical language, "out of a denser medium into a rarer." This follows, from the general fact that the illuminating and illuminated points are convertible, or that a ray can always return by the path of its arrival, so that the refraction of a ray out of any medium into air is performed according to the same rule of the sines, only reversing the terms of the proportion; or in other words, regarding what was the angle of incidence in the one case as that of refraction in the other and *vice versâ*. Numerically expressed, this reversal of the terms of a proportion, or ratio, is equivalent to inverting the numerator and denominator of the fraction expressing it,—so that, for instance, in the passage of light out of water into air, the "law of the sines" is expressed in the same general terms, but the "refractive index" (by which is meant the *number* expressing the proportion in question) has to be changed into its numerical reciprocal. In the case supposed, when light passes out of air into water, the proportion of the sines is that of 1336 to 1000, or almost exactly 4 to 3; and the "refractive index" is accordingly expressed by the fraction $\frac{4}{3}$, or the almost exactly equivalent decimal 1·336. In the reversed case, then, when the transmission is out of water into air, it will be $\frac{3}{4}$=0·75, or more precisely 0·749.

(26.) As a matter of experiment, it is found that between transparent media, or substances capable of being traversed by light, there exists a very wide diversity in this ratio of the sines of the two angles in question, or in the numerical values of the "refractive indices." Thus when light passes out of air into the less refractive species of plate-glass, the index instead of $\frac{4}{3}$ is $\frac{3}{2}$ or 1·5; into sulphur (which in its crystalline form is transparent), 2·0; and into diamond, or the mineral called octohedrite, 2·5. In fact, each particular transparent substance, solid, liquid, *or gaseous*, has its own peculiar, and, so to speak, *characteristic* index of refraction, which is found to stand in relation to its physical habitudes in many other respects, especially with its chemical composition, and its state of aggregation and density.

(27.) Even common air, in respect of a vacuum, has its refractive index—viz., 1·0003—the effect of which is perceived in the phænomenon of astronomical refraction, by which the sun or moon is rendered visible when actually sunk below the level of the true horizon.

(28.) From what is above stated, it is easy to see that when a ray is transmitted through a sheet or plate of any substance (as a window-glass) with parallel surfaces, its course after emergence will be parallel to its original direction, so that though displaced laterally, its direction in space is unchanged, which is the reason we see objects in their proper directions through a window. If the surface at which it emerges be not parallel to that through which it enters, this exact restoration of the original direction will not take place; and as we judge of the situation of an object only by the direction in which its light ultimately enters the eye, anything seen through a transparent substance whose surfaces are so inclined, will appear shifted in angular position. Any transparent substance so formed of polished plane surfaces inclined to each other, is called in optics "*a prism;*" and the angle at which the two planes in question meet, or would meet if extended, its "refractive angle." If such a prism—of glass, for instance—be held before the eye with its refractive angle vertical, and to the left, an object seen through it will appear deviated or shifted to the left of its true situation, the ray (as a slight consideration will show) being bent *towards the* thicker part of the prism. And thus by a very simple calculation, with which we shall not trouble our readers, from the angular amount of deviation caused by a prism of any medium whose refracting angle is measured, can the "refractive index" of that medium be ascertained.

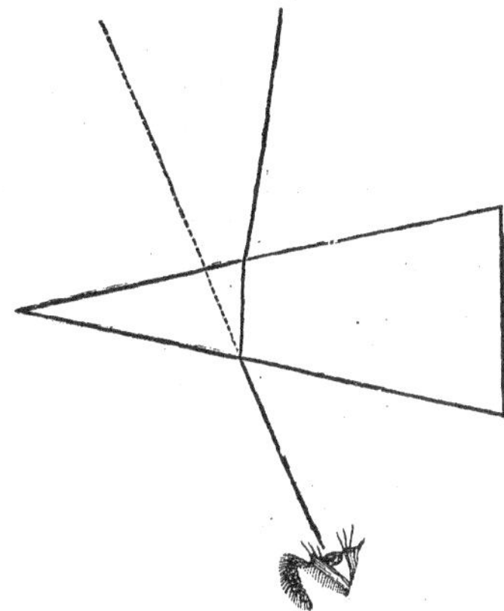

Fig. 2.

(29.) When refraction takes place out of *any one* transparent medium into *any other* in close and perfect contact with it—such contact as exists, for instance, between a fluid and a solid *that it wets,* or between two fluids of different specific gravities, which do not mix, resting the one on the other experiment shows that, so far as *the mere direction* of the refracted ray is concerned, it is the same as if the two media were separated by an exceedingly thin film of air. In that case, the same perpendicular being common to both surfaces at the point of contact, the angle of refraction out of the first medium is the same with that of incidence on the second. And from this it results that the proportion of the sine of internal incidence on the surface of the first to that of internal refraction at that of the second, or the "relative index of refraction," is constant for the same media, and is equal to the quotient of their respective *absolute* refractive indices. Thus, if the first medium be water, and the second be plate-glass, whose respective absolute indices are $\frac{4}{3}$ and $\frac{3}{2}$, the relative index, or that out of water into glass, will be $\frac{3}{2}$ divided by $\frac{4}{3}$ or $\frac{9}{8} = 1 \cdot 125$.

(30.) A very curious result follows from what has been said,—viz., that though light can pass out of a rarer medium into a denser, whatever be the obliquity of incidence, even when the incident ray but just as it were grazes the surface, yet the converse is not the case. For every denser medium, there is a limit of obliquity, beyond which transmission into a rarer cannot take place. The ray is *wholly reflected* without undergoing any diminution of brightness whatever; observing the same law of equality between the angles of incidence and reflexion, as in the case of ordinary reflexion on a mirror. The brightness of the reflexion, however, far surpasses anything that can be obtained from the most brilliant looking-glass or metallic mirror, being equal to that of the object directly seen. The effect is very striking, and is easily seen by immersing a small rod obliquely in a glass tumbler of water, and viewing the *under* surface of the water from below upwards at a moderate obliquity. The reflexion of the rod is seen without the smallest diminution of brightness. It is thus that fishes see the bottom of their pond reduplicated by internal reflexion on the distant parts of its surface. The *rationale* is simple enough. If two angles always have their sines in a fixed proportion, the greater may increase up to a right angle, but the less cannot: since the contrary would require the sine of the greater to exceed the radius of the circle.

(31.) Within this limit, when the angle of incidence is such as to admit of the transmission of the ray, the reflexion is less than total. The incident beam is subdivided; a part only is transmitted, the rest undergoes reflexion. The total amount of incident light is divided between them, but very unequally, and *the more* so *the less* the difference between the refractive indices of the media; or, in optical language, between their "refractive densities." Thus, when light passes at a perpendicular incidence out of air into water, only 2 per cent. of the whole incident beam is reflected; when into plate-glass, about 4 per cent., but when out of water into such glass, the amount of reflected light is less than ⅓ per cent. At oblique incidences, the reflexion is more copious, increasing in intensity as the obliquity increases, until the incident light but just grazes the surface.

(32.) The laws of reflexion and refraction being known, it is the part of geometry to follow them out in the several cases where light is incident on plane, spherical, or any other curved surfaces, reflecting or refracting, and thus to deduce the various theorems and propositions which the practical optician has need of for the construction of his mirrors, lenses, prisms, telescopes, and microscopes. All these, as beside our present purpose, we pretermit, confining ourselves entirely to the physical properties of light, and the theories which have been advanced for their explanation. This need not prevent us, however, from appealing to the effects produced by such instruments, especially such as are in most common use, and as can hardly be other than familiar to most of our readers, such as magnifying glasses (or lenses), telescopes, &c. It requires no knowledge of geometry, for instance, or any acquaintance with its application to theoretical optics, to enable any one to form a perfectly just conception of the mode in which the eye enables him to see, when his attention is called to a photographic picture, and he sees it impressed on its ground by the rays of light collected and brought to a focus by that assemblage of convex and concave lenses in a *camera obscura* which the photographer uses for the purpose. The dissection of an eye shows it *to be* such an assemblage, and the picture it produces may be actually seen at the back of the eye of an animal recently killed, by removing the opake leathery coat which envelopes it, and disclosing the retina. How the nerves of that tissue, indeed, convey to the mind *the perception* of colour and form, is, and will probably ever remain, a mystery; but is no more so in the case of vision than of any other of the senses; from which vision differs only in its transcendent refinement and the elaborate structure of that most wonderful of all optical instruments by which *form*, as well as colour and brightness, is brought within its range. The latter qualities are probably *perceived* by animals unprovided with eyes, such as the *proteus anguinus*, which inhabits dark caves, and whose delicate skin is evidently and painfully affected by the light; but to convey the perception of form, a picture must be produced, and in its own peculiar manner.

(33.) We are now prepared to understand the mode in which colour originates. This, to the ancients, was always a mystery. The light of the sun, and of ordinary daylight, which is only that of the sun dispersed and reflected backwards and forwards among the clouds, is white, or nearly so. Nevertheless, when we look through a red glass, or view a green leaf, it conveys to the mind the perception of those colours. How is this? If it be by light only that we see, and if that light convey to us *absolutely none* of the material elements of the bodies from which we receive it, how comes it that it excites in us such various and perfectly distinct sensations? The light itself must have either acquired or parted with something in its passage through or reflexion from the coloured body. Supposing, for instance, light to be a substance; it may have taken up some excessively minute portion of the object and introduced it to the direct contact of our nerves. In that case the sense of colour would be assimilated to those of taste or smell. Or it may have undergone *analysis*, and colour would then arise from a *deficiency* of something existing in the sun's light, and the *relative redundancy* of some other portion. In this view, light would be regarded, not as a simple, but a compound substance, or a mixture of so many simple ones as would suffice to explain all the observed differences of tint. On the other hand, if light be a movement, or an influence, we must admit in that movement or influence a similar capacity for analysis or composition, or else have recourse to some unknown modification of the one or the other, leaving the phænomenon as unexplained as before. There may, for instance, be a great variety of such movements, all *luminiferous*, but not all *alike;* and some may be destroyed, or some exaggerated, in the act of reflexion or transmission.

(34.) The key to this mystery, up to a certain point, was furnished by Newton, in his analysis of white light by prismatic refraction. A full account of the manner in which that analysis is performed, of the phænomena it presents, and of the nature and subdivisions of the "Prismatic spectrum," is given in our lecture on "The Sun," § 29, to which, to avoid repetition, we refer our readers. Let us, however, consider what kind of general theoretical interpretation we are entitled to put on this analysis. Now, the first and most obvious conclusion is, that the phænomenon we have to deal with, is not what in the accuracy of modern scientific language is understood by the term "analysis." It is the separation and redistribution (*according to degrees* of a certain quality common to all its elements—viz., that of REFRANGIBILITY) of a *mixture*, rather than the *di*alysis of a true *compound.* The simile by which we there illustrated it is so far exact. A glacier moraine might be redistributed by tidal action over the floor of the Ocean; the great blocks left *in situ*, or little moved—the smaller forming shingle, gravel beds, sandstones, or incoherent muddy deposits, with every possible intermediate gradation of size. But if in all this series any particular size were found entirely and universally deficient, throughout the whole series of formations traceable to that source, we should conclude, not that a mass of that size is an impossibility *in rerum naturâ*, but that owing to some unknown cause in the nature of a previous sifting, every pebble or grain of that size had been already separated, or otherwise arrested *in limine*, and might expect elsewhere to find it in the case of some other series of geological formations. So it is with the sun's light. Certain definite and marked degrees of *refrangibility* are wanting in its spectrum, indicated by the dark lines which cross it. But if absent in solar light, they exist in the light of flames, and of other luminous sources, which in their turn are again deficient in other degrees which yet abound in the solar rays. Refrangibility, then, taken as a property of light generally, *is* a quality susceptible of indefinite *gradation*, from the one extreme of the spectrum to the other.

(35.) If we limit our consideration to some one medium—glass, for instance—we find each particular degree of refrangibility associated, first, with a determinate and invariable index of refraction, which determines its place in the spectrum by determining the amount of deflexion it shall undergo in passing through the prism; and, secondly, with an equally determinate and invariable tint in the scale of "prismatic colour," the red corresponding to the least and the violet to the greatest refractive index. The truth of these propositions is easily tested on any one ray of the spectrum insulated from the rest by intercepting all the others. The ray so insulated, whatever its tint, is no longer separated or "dispersed" by subsequent refraction into a new spectrum. It preserves its tint unaltered, and conforms to the "rule of the sines" in its flexure, as if no other colour or refrangibility existed. Hence we might be led to conclude, as Newton himself did, that between these two qualities—refrangibility and colour—an absolute and invariable connexion exists. This, however, is not the case. The propositions in question cannot be generalized. When different media are examined, we find that not only does the same colour correspond to different degrees of refrangibility, or to different *absolute* values of the refractive index in each, but that the same *change* of colour does not correspond in different media to the same *proportionate* change of the refractive index; and that, in short, taking the "scale of colour" in all its gradations, from red, through orange, yellow, green, blue, and indigo, to the least perceptible violet, and that feeble tint beyond the violet which can hardly be called a colour, but which is most nearly expressed by the term *lavender*, as a guide,—each particular medium distributes these rays through its spectrum, though always in the same *order* of succession, yet in other respects according to a law peculiar to itself: thus indicating both a total amount of *dispersion* and a *scale of action* dependent on the physical properties of the medium, and in some sort as it were personal to each. This power which a transparent medium has of separating the differently-coloured rays and spreading them over an angular space greater or less in proportion to the total deviation of some one ray, taken as a standard, from its former course, is called in optics the "dispersive power" of the medium. It differs very widely in different media, and in consequence, the lengths of the spectra which they produce, corresponding to one and the same mean or average refraction, differ accordingly. Thus, for example, the total lengths of the spectra produced by prisms of fluorspar, water, diamond, flint glass, and oil of cassia (the mean refractions being the same), are to each other in the proportions of the numbers 22, 35, 38, 48, and 139.

(36.) This quality of dispersion stands in very distinct relation to the chemical constitution of the refracting medium. Thus it is found that all the compounds of lead, whether in liquid solution, natural or artificial crystals, or glasses into which that metal enters largely, possess very high dispersive powers; while those into which strontia enters exhibit remarkably low ones. It is on this property of lead that the formation of highly dispersive glasses, to imitate the brilliant colours of gems, and to give the vivid prismatic colours of the pendants of chandeliers by candle-light, depends, as well as that far more important application which, by the combination of two glasses of different dispersive powers, the one containing lead, the other none, enables the optician to effect refraction without producing colour, and so to construct that admirable instrument, *the achromatic telescope.*

(37.) Not only are the *total lengths* of the spectra produced by different media different for the same mean amount of refraction, but within those lengths the distribution of the several colours differs, the spaces occupied by the several tints differing very considerably in proportion to each other and to the whole. Thus, in the spectrum formed by flint-glass, and most other of the highly dispersive media, the green is situated nearer to the red than to the violet end of the spectrum, while in that formed by muriatic ("hydrochloric") acid the reverse is the case.

(38.) By the reunion of all the coloured prismatic rays (which may be effected by an equal and contrary refraction of the whole spectrum through a prism of the same material reversely placed), white light is reproduced. And hence we conclude that *colour is not a superinduced but an inherent quality of the luminous rays.* Again, if we exclude from this reunion any portion of the spectrum, the reconstituted beam is coloured: and if the rays so excluded be not *extinguished,* but diverted aside, and themselves collected and reunited into another and separate beam (which may easily be effected, with a little management, by one skilled in experimental optics), this will also be coloured, but with a tint *complementary* to that of the first. Between the tints so arising is always found to prevail that beautiful and, so to speak, *harmonious contrast* which is so effective in the ornamental arts, where one colour is said to *set-off* another, or show it to the greatest advantage. Thus, crimson or pink is complementary to green, scarlet or orange to blue, yellow to purple, &c. The relation to each other of these complementary colours is curiously and strikingly illustrated by the spontaneous production within the eye itself of the tint complementary to any vivid colour, which takes place when, after gazing steadfastly on an area so coloured, on a white ground, and strongly illuminated, the gaze is suddenly transferred to a uniformly white surface. There is seen on it, though only for a few moments, a picture or optical image of an area similar in form and size, but tinted with the complementary hue, which fades quickly away. This curious and beautiful experiment, which requires no apparatus to exhibit, and which any one may try in a moment, is exceedingly illustrative of the mode in which the *sensation* of colour is produced. It proves that, in the nervous tissue which receives and feels the picture within the eye, there are nerves individually and exclusively sensitive to each of the coloured rays, or at all events to each of those primary colours (if such there be) by whose mixture all colours are compounded. When white light falls on a portion of the retina wholly or partially deadened or fatigued by the excitement of the nerves appropriate to one set of rays, the sensibility of the others being left unexhausted; that other portion will be for a time proportionably more sensitive to the remaining rays: so that under the stimulus of white light an undue preponderance is temporarily given to their influence, and the sensation of the complementary tint is conveyed to the mind. This is only one of innumerable instances of the wonderful adaptation of that most astonishing organ to the performance of its office of conveying to us information not only of the forms and situations of objects, but of all that multitude of their physical properties which stand in relation to colour, both those which ordinary experience teaches and which science reveals.

(39.) Lastly, by thus reuniting into one beam rays going to form distant portions of the spectrum, and excluding the rest, we find that it is possible to produce a compound beam which shall excite directly in the eye, or illuminate a screen with any one of the innumerable varieties of tint which we observe in nature; and what is especially remarkable, the same tint, or one undistinguishable from it *to ordinary eyes*, is producible by very different combinations of the prismatic rays; while yet there exist individuals, and these not unfrequent, who are perfectly capable of discriminating (in many cases) between such compound tints, and who even declare them to be widely different. To such cases of what is called, though improperly, "colour-blindness," we shall presently have occasion to recur.

(40.) The consideration of these facts has given rise to a speculation which, if not demonstrable, has at least a high degree of plausibility, and which, at all events, has never yet been *dis*proved,—viz., that there is no real connexion between COLOUR and REFRANGIBILITY, but that there exist three inherently distinct *species of light*, each competent *per se* to excite the sensation of one of three PRIMARY COLOURS, by whose mixture all compound tints are produced, white consisting of their totality, and black being the exponent of their entire absence. That, moreover, each of them has a spectrum of its own, over the whole length of which it is distributed according to its own peculiar *law of intensity*, and from whose superposition on the same ground results the *prismatic* spectrum, coloured as we see it. The annexed figure will convey a better conception of

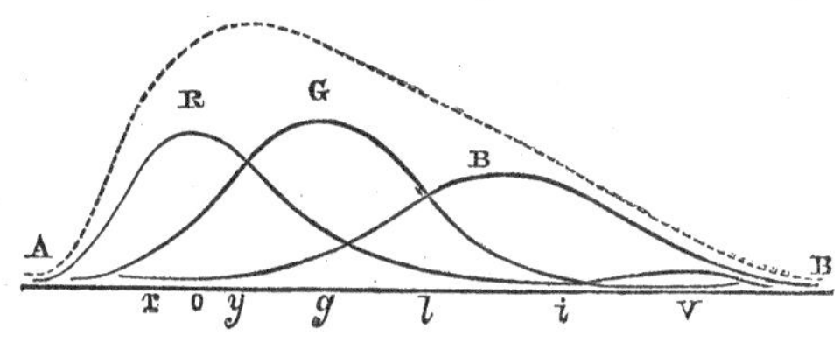

Fig. 3.

this than any lengthened description, where A B represents the length of the total spectrum wherewith each of the three is co-extensive, and where the curved lines marked R, G, B, severally express, by the height to which they rise on any one point in A B, the intensity in its own spectrum of each of the primary colours; while the dotted curve, whose *ordinate* or height corresponding to any point is the sum of those of the other curves, will of course express the joint intensity or degree of illumination in the visible spectrum.

(41.) In this view of the subject, the prismatic colours, with the exception of the extreme red, are all more or less mixed tints, and this agrees well with its general aspect, in which the red and indigo-blue are the only full and pure tints, the green being by no means a *saturated* or full green, and the violet having a strong dash of purplish-red in it.

(42.) The three primary colours assumed in the above figure are red, green, and blue, each in its highest degree of purity and *undilution*, for it will be readily apprehended that while the admixture of any one, in however small a proportion, will produce a rich though a mixed tint, that of *both* the others tends to dilution. The only three colours which answer all the experimental conditions, are these three. This may seem contrary to the experience of the artist, who from his habitual practice in mixing the colours he uses (all of them without exception compound tints), would name yellow, in place of green, as the intermediate primary. The reason is obvious. In all the yellows which he uses there is a large admixture of red with green, and in all his blues more or less green. When mixed, then, there is sure to be a preponderance of green, while the red goes to neutralize a portion of the other two, and so to dilute the outstanding green. On the other hand, *the direct mixture of the prismatic yellow and blue, in whatever proportions, can no-how be made to produce green*, as Professor Maxwell's, M. Helmholz's, and my own experiments* have distinctly proved; while that of the prismatic green and red *does* produce yellow. This will be better understood when we come to speak of the *absorption* of coloured light.

(43.) Since at each point of a compound spectrum so constituted, all the three primary elements, in whatever proportion mixed, have one and the same degree of refrangibility, it is evident that the compound tint

* See "Notices of the Royal Society," vol. x. p. 52.

arising from their mixture cannot be separated by any subsequent refraction into its components.

(44.) In persons who are what is called "colour-blind," the eye is sensible to all the rays of the prismatic spectrum *as light*, though even in that respect the red rays appear comparatively *deficient in power* to stimulate the nerves of vision, so that all colours, into which a large proportional admixture of primary red enters, are described by them as sombre tints. But besides this, two of the primary coloured rays, the red and the green, appear to excite in their nerves sensations of *colour* nearly or exactly similar. Their vision is therefore, in fact, *dichromic;* all their compound colours are resolvable into two elements only instead of three. Red they do not distinguish from green. The scarlet coat of the soldier and the turf on which he is exercised—the ripe cherries and the green leaves among which they hang—are to them undistinguishable by colour, though from constantly hearing them so spoken of, they habitually speak of the fruit as red and the leaves as green. Their sensation of blue is *probably* the same as in normal vision; though whether that excited by their other colour, be such as a normal-eyed person would call red, yellow, green, or something quite different from either, we have no means of ascertaining, nor can they give us any information. The face of nature must appear, however, to them far inferior in splendour and variety to that which we behold; and if there be, as is asserted, here and there an individual totally destitute of the sensation of difference of colour, it must present to his eyes what we should be disposed to call a hideous monotony—light and shade only revealing the forms of objects as in an engraving. Yet what we never knew we never miss. There may, and not improbably do, exist beings in other spheres, if not here on earth, whose vision is sensitive to those rays of the spectrum which extend far beyond the violet or its *lavender* prolongation, and which we know at present only by their powerful photographic activity, and by their agency in producing that singular species of phosphorescence in certain media to which Professor Stokes has given the name of Fluorescence. By these properties, the solar spectrum is proved to be prolonged far beyond its visible limits at its most refracted extremity; as it is by *other invisible rays* OF HEAT, which have been traced up to nearly an equal distance beyond the extreme red in the opposite direction.* All, however, whether of heat or chemical influence, conform each for itself, and according to its own special "refractive index," to the same general *law of the sines*, as well as to every other of those singular and complicated relations of the luminous rays, we shall hereafter have to describe; and both the one and the other extending into and thinning out as it were in the luminous region, just as we have described the spectra of the primary colours into those of each other. Such, and so wondrously complex a compound is a sunbeam!

* See my paper in the Phil. Trans. R. S. 1842, "On the Action of the Solar Rays on Vegetable Colours."

(45.) The analysis into its prismatic elements of the colour of any natural object, is readily performed by examining through the refracting angle of a prism of perfectly colourless glass a rectilinear band or strip of the colour to be analysed, so narrow as to have scarcely any apparent breadth, and to appear as little more than a coloured line. Placing this on a perfectly black ground, parallel to the refracting edge of the prism, and illuminating it as strongly as possible, it will be seen dilated into a spectrum, or broad riband of colour, exhibiting of course those coloured rays only which belong to the composition of the tint examined. An exceedingly convenient arrangement for this purpose is to fasten across one end of a hollow square tube of metal or pasteboard blackened within, of about an inch square and twelve or fourteen inches long, a metal plate having in it a very narrow slit parallel to one side, quite straight, and very cleanly and sharply cut. At the other end within the tube is to be fixed a small prism of highly dispersive colourless flint glass, having its refracting angle parallel to the slit, and so placed that when the tube is directed to the sky, or rather to a white cloud, the slit shall be seen dilated into a clear and distinct prismatic spectrum. In this of course all the prismatic colours will be seen in their due order. But if, instead of this, any coloured object—as the leaf of a flower, for instance, or a coloured paper, strongly illuminated by direct sunshine (if necessary, concentred on it by a lens, so, however, as not to *scorch* the object by the heat of its focus),—be placed so near to the slit as completely to occupy its whole area and suffer no ray to enter it which does not come from some part of the coloured surface; the spectrum will be seen deficient in all those rays which the object does not reflect, and which belong to its complementary colour. The use of this little instrument, at once simple, portable, and inexpensive, will be found to afford an inexhaustible source of amusement and interest. To the florist, on a bright sunny morning, the analysis of the tints of flowers and leaves, or the hues of a butterfly's wing, and of every variety of coloured object;—to the water-colour painter, the study of the prismatic composition of his (so fancied) simple washes of colour and the effects of their mixture and superposition;—to the oil painter, that of the various brilliantly coloured powders which mixed with oil form the material of his artistic creations, are all replete with interest and instruction.

(46.) If instead of a reflected colour we would examine a transmitted one, as in the case of a coloured glass, or some natural transparent coloured product,—if in the form of a plate or lamina, it may be laid over the slit, and when directed to any bright white light (as that of a white cloud), its spectrum will be exhibited—if a coloured flame, the slit may be placed close to it, but if a liquid, it will be preferable to make it its own prism by enclosing it in a hollow prism formed of plates of glass cemented together, when the differences arising from difference of the thickness of the medium traversed by the refracted rays will be more easily studied.

(47.) The colours of transparent media—such as

coloured glasses, crystals, resins, and liquids—depend upon the greater or less facility with which the several coloured rays are transmitted through their substance. There is no medium known, not even air or the purest water, which allows all the coloured rays to pass through it with equal facility. Independent of the partial reflexion which takes place at the surfaces of entry and emergence, a portion greater or less according to the nature of the medium, is always stifled, or as it is called in optical language, "*absorbed:*" and this absorptive action is exerted unequally on the differently refrangible rays; so that when a beam of white light is incident on any such medium, it will be found at its emergence deficient in some one or more of the elements of colour, and will therefore have a tint complementary to that of the absorbed portion. Supposing, as is most probable in itself, and agrees with the general tenor of the facts, that an *equal per-centage* of the light of any specified colour which arrives at any depth within the medium is absorbed in traversing an equal *additional* thickness of it, the intensity of the coloured ray so circumstanced would diminish in geometrical, as the thickness traversed increases in arithmetical progression. The more absorbable any prismatic colour, then, the more quickly will it become so much reduced in proportion to the rest as to exercise no perceptible colorific action on the eye. And thus it is found that in looking through different thicknesses of one and the same coloured glass or liquid, the tint does not merely become *deeper and fuller*, but changes its character. Thus a solution of sap-green, or of muriate of chromium, in small thicknesses is green—in great ones red; tincture of violets, and that species of rich blue glass which is coloured with cobalt, in like manner are red when we look through a great thickness, but beautifully blue when thin; and so in a multitude of other cases. Those who paint in water colours are well aware of what importance it is to effect the tint they aim at by a single wash of their colour. A second application of the very same liquid, after allowing the first to dry, does not simply heighten the colour, *but changes the tint*, a circumstance which those who practise that fascinating art will do well to bear in mind.

(48.) When white light is transmitted successively through two or more coloured media whose scales of absorption differ materially, the residual beam, or that which struggles through after passing their successive ordeal, will consist of those rays only whose transmission is favoured by all the media. Hence it will follow, first, that the final *tint*, or that of the beam ultimately emergent, will most probably be very different not only from that exhibited by either of them separately, but from that which *might be expected to arise* from a union or blending of their tints, and which *would arise* were we to *unite together* distinct luminous beams having those tints; and, secondly, that all such successive transmissions tend to produce sombre tints, and ultimately complete blackness; inasmuch as each successive transmission destroys (or absorbs) a greater or less proportion of the total illuminating power of the original beam. Thus when colour is produced on white paper by the laying on of successive washes of *different* transparent colours, the tendency is to produce, first, a tint very remote from that *expected* to result from their union; and secondly, becoming more and more muddy and sombre, the greater the number of such heterogeneous layers of colour. Hence the maxim in water-colour painting, to secure brilliancy by using only a single wash of colour, if possible, to produce the required effect. The painter should never forget that his notion of colour (as compared with that of the photologist) is *a negative one.* He operates solely by the *destruction* of light, and his aim should always be to destroy as little as possible. His direct action (unknown to himself) is upon the tint complementary to that which he aims at producing.

(49.) Each particular coloured medium has its own peculiar and specific scale of absorptive action, differing *inter se* in the most singular and capricious manner. In many, indeed in most cases, the spectrum viewed through such a thickness as to give a strong colour to common daylight, in place of being seen as a continuous band of graduating colour, is broken up into distinct coloured spaces, more or less intense, and more or less well-defined, separated by dark intervals. This is particularly the case with coloured gases or vapours. Thus the red vapour of nitrous gas, especially when its absorptive action is intensified by heat, breaks up the spectrum into a succession of narrow spaces, alternately dark and bright, from one end to the other.

(50.) When coloured flames are examined with such a "spectroscope" as above described, the phænomena are no less varied, and in the highest degree characteristic. The presence in the flame of *each particular chemical element* determines the presence in its light of some one or more coloured rays of *definite* refrangibility and colour, producing often in its spectrum the appearance of a definite line of coloured light out of all proportion brighter than the rest. Thus the presence of soda in any flaming body is characterized by a narrow and exceedingly vivid line of yellow light. So completely characteristic are these lines of the chemical elements to which they bear relation, that no less than four new metals, Thallium, Rubidium, Cæsium, and Indium owe their first discovery to the observation of definite spectral lines of their appropriate colour, produced by their presence in quantities too minute to be rendered sensible in any other manner.*

(51.) It is impossible in the compass of a lecture like the present, to do more than notice with extreme brevity these remarkable classes of phænomena, and that only as bearing upon the general object we have in view. They prove in the most convincing manner the close and intimate relation in which LIGHT stands to MATTER. It enters into the interior of the hardest and least penetrable bodies, and thereout brings us information of an

* In reference to what is now called "Spectrum Analysis," in a chemical point of view, I may be here allowed to call attention to a passage in my "Treatise on Light," published in 1827 (Encyc. Metrop., vol. iv.):—"The colours thus communicated by the different bases to flame, afford in many cases a ready and neat way of detecting extremely minute quantities of them."—Article, "Light," § 524.)

almost infinite variety of particulars as to their intimate nature and constitution (and, as we shall see further on, of their internal structure, and the mechanism by which they are held together as bodies), which by no other means we can obtain: information which at present we are only imperfectly able to interpret, but whose import, from year to year, and almost from week to week, is becoming better understood. Its language in this respect bears no distant similitude to that of a series of ancient inscriptions in some unknown tongue and character. A single sentence once developed by some happy and unmistakable concurrence of evidence, affords a clue to others, which in their turn become the stepping-stones of further progress. By the one are revealed the histories of ages long buried in oblivion, and of the phases of human thought and action under circumstances bearing little analogy to anything we now see around us: by the other we are admitted a step nearer to the perception of the intimate working of those powers which maintain the material universe as it stands, and the laws they observe.

LECTURE VII.

ON LIGHT.

PART II.—THEORIES OF LIGHT—INTERFERENCES—DIFFRACTION.

TWO theories only, entitled to any consideration as rational and intelligible explanations of the phænomena of Light, have been advanced—the one proposed by Sir Isaac Newton, commonly known as the "Corpuscular;" the other by Christian Huyghens, as the "Undulatory" theory. According to the former, light consists in "Corpuscules," or excessively minute material particles darted out in all directions from the luminous body, in virtue of some violent repulsive power, or other energetic form of internal action, acting under such circumstances, and under such laws, as to give them all the same *initial* velocity which they retain unchanged in their progress through space, as well as their initial direction according to the general laws of motion (to all which they implicitly conform), until they meet with some material body by whose action their course is changed. All this, and all subsequent changes of direction and velocity, are held, on this theory, to be effected by attractive or repulsive powers resident in the bodies on which the light-corpuscules fall (or, which comes to the same thing, in the corpuscules themselves), and from which they are either reflected, if the repulsive powers be too strong to permit their penetration; or in which they are refracted, if they are able to enter and make their way among the particles of the refracting body. Colour, according to this theory, is accounted for by specific diversity among the luminous particles; and difference of refrangibility, by differences in the intrinsic energy of the acting forces as determined by the specific nature of the molecules, or, which comes to the same, by a difference of proportion between their *moving force* and their *inertia*. This is one of the many weak points of the theory. It runs counter to the only analogy which the observation of nature furnishes. It is as if the sun should be supposed to attract a planet of lead and one of cork with different *accelerating* forces; or as if, here on earth, a lump of platina and a lump of iron should be supposed to acquire different velocities in falling through the same space. It runs counter, too, to the original assumption, that when first emitted from a luminous body, in their passage through empty space, all the coloured particles move with equal velocities, and *have* therefore been *equally accelerated* by the emitting forces. That they *do* so, we know from astronomical observation. The *Aberration* of all the coloured rays is the same. Were it not so, every star seen through a highly magnifying telescope ought to appear as drawn out into a short, coloured *spectrum* in a certain definite direction. Light requires forty-two minutes to reach the earth from Jupiter at its mean distance. Supposing the rays of one end of the spectrum—the violet, for instance—to travel faster than those at the other (the red), a satellite undergoing eclipse by immersion in the shadow of the planet ought to change colour before extinction, from white to red—the last-emitted red rays lagging behind the violet on their journey to the earth; while at its reappearance a blue colour ought to be first perceptible.

(53.) Among the stars are many which vary periodically in brightness, and some of them undergo complete extinction. As light takes *several years* to travel from the stars, the difference in the times of arrival for any sensible difference of velocity would amount to many days, and would be quite sufficient to tinge the disappearing and reappearing star with the hues belonging to opposite ends of the spectrum. No such thing, however, is observed. Most of them retain their whiteness; and though some *do* assume a deep-red colour when undergoing extinction, or when at their minimum of splendour, it is *not* changed to blue at their reappearance, or on their commencing augmentation of brightness.

(54.) The reflexion and refraction of light are, as we have stated, accounted for on this theory by supposing the particles of all material bodies, besides the attractive force of gravitation, to be endowed with other forces, both attractive and repulsive—the latter extending to a greater distance than the former, so as to constitute an attractive and a repulsive sphere one within the other—the particles of light being repelled while passing through the outer or repulsive sphere, and attracted when arrived within the internal or attractive one. These forces are supposed immensely energetic, and to decrease with such excessive rapidity as to be absolutely insensible at any, the very smallest, distance appretiable to our senses. In virtue of this repulsive force, the surface of any material body may be conceived as coated (metaphorically speaking) with a film of repulsive power, off which, as from an elastic cushion, the luminous particles may be imagined to rebound: in which case, according to the known laws of elastic rebound, the angle of reflexion (perfect elasticity being supposed) would be equal to that of incidence, and the velocities before and after reflexion equal.

(55.) Reflexion, then, is easily and readily explained on this theory. In fact, it is explained *too well.* For it will be at once asked, how, on such suppositions, there can be such a thing as *partial* reflexion. Since all the luminous particles of a ray arrive at (suppose) a plane surface in the same direction and with the same velocity; whatever happens to one, *the repulsive force being the same*, must happen to all. This is another weak point of the corpuscular theory; and to escape from the difficulty so created, it becomes necessary to supplement the original hypothesis of *luminous particles* with another, converting those particles into mechanisms of a peculiar nature, of which the simplest conception that can be formed is to suppose them as it were *minute magnets* having attractive and repulsive poles, and during their progress through space revolving round their own centres about axes not coincident with the direction of their motion. Under such circumstances it is clear that some might arrive at the reflecting surface with the attractive pole foremost—others with the repulsive. The former would be attracted, and escape the reflective action; the latter repelled, and therefore subjected to it. Or, without making any supposition as to the sort of mechanism by which such a result might be attained, we might content ourselves with assuming, as Newton (the framer of this hypothesis) did, that the particles of light, throughout their whole progress through space, pass periodically through a succession of alternating physical states—or, as he called them, "fits"—"of easy reflexion and easy transmission:" the only objection to such a form of statement being, that it conveys no clear *physical* conception to the mind.

(56.) The particles so escaping reflexion are conceived to have penetrated within the limit of the repulsive, and to have entered that of the attractive forces, while yet at some inconceivably minute distance *outside* of the actual surface of the medium. Their movement of approach therefore to the surface is accelerated by the attractive force whose resultant direction is perpendicular to the surface, and when they have arrived *within* the medium so far that all further action ceases (by the counteraction of equal and opposite forces on all sides) each of them will have undergone the total amount of acceleration due to the attractive force—*in the direction of that force*, *i.e.*, at right angles to the surface. Its velocity estimated in this direction will therefore be greater within the medium than without—while that parallel to the surface remains unchanged: the force in that direction being *nil.* The direction of the motion therefore will be more highly inclined to the surface within the medium than without, in the same manner and for the very same reason, that the path of a projectile shot obliquely downwards from the top of a hill makes a greater angle with the horizon when it reaches the ground than it did in the commencement of its descent. And the conclusion, on strict dynamical principles, is the same in both cases. Supposing the initial velocity of projection the same, the sines of the angles made by the direction of the motion *with the* vertical or perpendicular to the surface, at the beginning, and at the end of the descent (*i.e.*, in the case of light, those of the angles of incidence and refraction), will be to each other in an invariable proportion, the total *height* of the descent being the same. Thus we see that the law of refraction is satisfactorily accounted for, on the corpuscular hypothesis; and that, on that theory, the velocity is greater in the interior of a refracting medium than in empty space; and the more so, the greater the refractive power.

(57.) Let us now see in what sort of conclusion we are landed as to the intensity of the forces we have pressed into our service. To consider only the reflective force, we have this to guide us—that, supposing the incidence perpendicular, and the light therefore reflected back by the path of its arrival, that force must have been sufficiently great to destroy the whole velocity of the luminous particle, and to generate an equal one in the opposite direction, in the time occupied by the particle in traversing forwards and backwards the thickness of our stratum of reflecting force. Now the velocity of light, as we have seen, is 186,000 miles per second. To destroy and reproduce this velocity in a projectile shot directly upwards, by the force of gravity on the earth, supposed uniform or undiminished by distance, would require its action to be continued for 706 days, or very nearly two years, while the same effect has to be produced by the reflecting force (also supposed uniform), in that inappretiable instant of time *in which* the act of reflection is performed—a time which would be extravagantly overrated at the *billionth** part of a second. After this we need hardly trouble our readers with any estimation of the intensity of the refracting forces. The sturdiest philosophy may fairly be staggered at such a postulate as the foundation of a physical theory.

(58.) According to the "undulatory theory" light consists in an undulatory or vibratory *movement* propagated through an elastic medium pervading all space, not even excepting what is occupied, or seems to be occupied, by what we call material bodies—that is, such as have *weight*, and which, to us, constitute the visible and tangible universe of things. It therefore resembles sound, which is not a travelling entity, but a propagated motion in the air, analogous to the tremulous movement which runs from

* A billion is a million times a million. The French *milliard* is a thousand millions.

end to end of a stretched cord, or to the waves which *appear* to travel along the surface of water; though in reality such a wave is only an *advancing form*, the real movement of the watery particles being vertically up and down. Colour in this view of the subject is analogous to *tone*, or *pitch*, in music (if it be supposed to depend solely on refrangibility). As the *frequency* of the vibrations which reach the ear from a sounding-string determines the pitch of the musical note it yields, so the frequency of the undulations of this elastic medium or luminiferous "*ether*," as it is called, determines to the nerves of the eye the colour of the light. *Or* in that view of colour which considers all but three primary hues composite, it must on this theory be assimilated to a difference analogous to *quality* in a musical tone—as, for instance, between the sounds of a violin, a flute, and a trumpet, only much more decided and strongly characterized.

(59.) As sound spreads through the air with equal rapidity in all directions, and may be considered as propagated from its origin as a spherical shell continually enlarging, so in this theory must light be regarded as the movement of a WAVE in the ether, running out spherically in all directions from the luminous point, whose situation with respect to the eye, or to any other point on which the wave may strike, is judged of as the centre of the sphere—*i.e.*, as lying in a line perpendicular to its surface. A *ray* of light then, in this theory, is a purely imaginary line from such point, perpendicular to the general surface, or *front* of the wave, and has no other meaning. The *wave*, not the *ray*, is the primary object of contemplation. If the point where the luminous excitement originates be near, the perpendiculars from it to the wave-surface diverge conically; but if so far remote that the portion of that surface at the eye may be regarded as sensibly plane, they are to all sense parallel, as in the case of light emanating from the stars or the sun.

(60.) The reflection of light in this theory is in exact analogy with that of any other undulatory movement. We cannot see the waves of sound, but those on smooth water are easily followed and their reflexion made matter of ocular inspection. Drop a small pebble into still water, and a wave will be seen to spread out in an enlarging ring. Let this be done near the perpendicular and smooth side of any large tank or pond, or near a board held vertically in the water, and the ring will be seen on reaching the board to be reflected, and will thence spread back over the surface, still enlarging, as the segment of another ring whose centre might be supposed as far on the land side of the reflecting surface as the place where the pebble was dropped was in reality on the water side. If several pebbles be dropped in succession, or a regular up-and-down movement given to the water at that point, a continued series of circular waves will be generated and reflected,—the reflected waves running out and intersecting the direct exactly as if they originated in two distinct centres. What in water is seen to be a reflected wave, in air we recognize as an *echo*. And in the fact that a sound, though partially reflected as such from a window, a board partition, or a wall, is heard, though with diminished intensity, on the other side,—we have the analogue to the partial reflexion of a beam of light at a transparent surface; and on the other hand, in the deadening of sound in passing through woolly or puffy substances, while it is transmitted with exceeding sharpness and distinctness through compact solids or through water, we have the parallel to the absorption of light in some media, and its copious transmission through others.

(61.) The explanation of refraction on the undulatory theory is exceedingly simple. Suppose a plane wave to sweep obliquely along the surface B E of a medium capable of propagating within it the luminiferous undulation, and let it be supposed at equal intervals of time

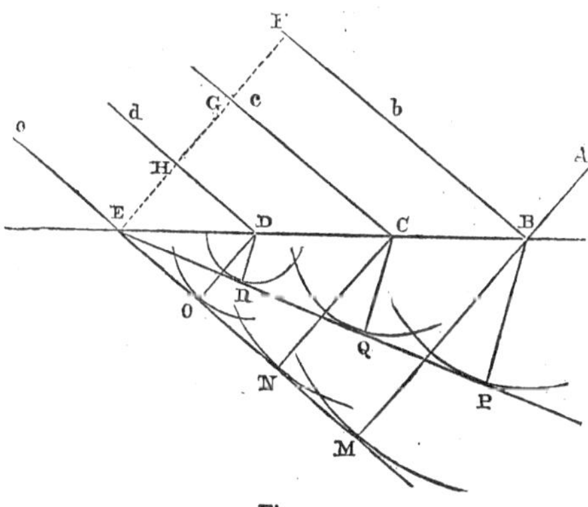

Fig. 4.

(successive seconds, for instance) to assume successive positions B b, C c, D d, E e, arriving in succession at equidistant points B C D E of the surface. So soon as any point in the surface is struck by the wave it will be set in undulatory motion and propagate from it a movement which will run out spherically from that point in all directions with such (uniform) velocity as belongs to the luminous undulation *in* the medium. When, therefore, the wave has reached the position E e, E will just have begun to move; the internal wave propagated from D will have travelled during one second, from C two, from B three seconds,—and the motion, in virtue of these, respectively, will have extended to the surfaces of spheres about those points as centres, having radii in the proportions 1, 2, 3, so that a plane passing through E, which touches one of them, will touch them all, and the same is true for all points intermediate between these. Such a plane will define *the limit up to which the movement has reached* within the medium when the exterior wave has the position E e, and will, therefore, be *the front of a plane wave* advancing within it. If the velocity of the undulation within the medium be the same as without, D O, C N, B M, the radii of our spheres will be equal to E H, E G, E F, the spaces run over in one, two, three seconds outside, and the touching plane E O N M will evidently be a continuation of the exterior plane wave e E. In this case, then, there is no *refraction*, the direction of the interior ray B M being the same as A B, perpendicular to the exterior wave. But suppose the velocity within the medium less than that without. In that case the radii of our spheres D R, C Q, B P, will be less than D O, C N, B M, and in a constant proportion. The plane E P touching them all then, or the front of the interior wave will be inclined at a less angle B E P to the surface than B E M, or its equal E B F,—and the sines of these angles to a common radius E B are evidently in the proportion to each other of P B to B M, or of the *velocity of light* in the *medium to the velocity out of it.* Now, as the *ray* is perpendicular to the *wave*, the inclination of the latter to the surface is the same as that of the former to the perpendicular, and thus these angles are respectively identical with those of refraction and incidence.

(62.) To such of our readers as may find a difficulty in following out this reasoning, the following familiar illustration will convey a full conception of its principle. Imagine a line of soldiers in march across a tract of country divided by a straight boundary line into two regions, the one smooth, level, and well adapted for marching, the other difficult, rough, and in which from its nature the same progress cannot be made in the same time. Suppose, moreover, their line of front oblique to the line of demarcation between the two regions, so that the men shall arrive at it in succession, and not simultaneously. Each man, then, from the moment he has stepped across this line, will find himself unable to make the same progress as before. He will be therefore unable to keep line with that part of the troop which is still on the better ground, but must of necessity lag behind; and that, by the greater space, the longer he travels. Since each man on his reaching the line of division experiences the same difficulty: if they will not break line and straggle, but persist in still marching *in line* and keeping up their connexion, it will follow of necessity that the front of *their* line must to a certain extent fall back and make an obtuse angle at its point of junction with that of the unimpeded line. Thus, in our figure, B E will represent the line of division between the two regions, B b the advancing front of the troop when the first man arrives at that line, E e that of the portion still on the good ground after some time elapsed, and E P that of the other portion who have been unable to keep up to the same rate of march. And as the necessity of keeping step and not crossing each other's line of march will oblige each man to step out *right in front* (*i.e.*, at right angles to the new frontage), the progress, B P, made by the first man after crossing the line, will be perpendicular to E P, and will be to what he would have made (B M) had it not been for the retardation, in the proportion of his new to his former velocity of march.

(63.) Thus then we see that when light passes (in this theory) out of what is called a rarer medium into a denser, or when the angle of refraction is less than that of incidence, the velocity of propagation of the undulatory movement is *diminished*, while on the corpuscular doctrine it is increased, and *vice versâ*. Thus, too, we see that on the undulatory hypothesis the connexion between refrangibility and velocity within the refracting medium *is immediate and absolute*, and consequently that it being certain, as we have shown, that light of all refrangibilities travels equally fast in *what we call* empty space (*i.e.*, through the ether alone), it follows with equal certainty that in material media the more refrangible rays are propagated slower than the less so; and all, more slowly than in free space. In other words, this amounts to supposing the elastic force of the ether either to be enfeebled in the interior of material bodies, or else that the movements of its particles are *in some way or other* clogged or burthened by some sort of connexion with or adhesion to the material molecules among which they are disseminated, and *that* more for the more refrangible rays than for the less so.

(64.) Until lately this difference of velocity between the differently refrangible rays had always been considered an insuperable obstacle to the admission of the undulatory hypothesis. All sounds of whatever pitch (it was contended) travel equally fast in one and the same elastic medium. The profounder researches of later mathematicians, however, have shown that this conclusion is not absolute, and that on certain suppositions which are not altogether inadmissible in respect of the vibrations of light, the difference is not contradictory to strict dynamical laws.

(65.) As we have attempted to form an estimate of the intensity of the forces required to account for observed facts on the corpuscular hypothesis, let us now attempt a parallel estimate on the undulatory. And here the way is equally open and obvious. Starting with the observed facts, that sound travels in air at the rate of 1090 feet per second, while light is propagated through the ether 186,000 miles in the same time (that is to say, 901,000 times as fast), we are enabled to say how many fold the elastic force of the air, or its resistance to compression, would require to be increased *in proportion to the inertia of its molecules*, to give rise to an equally rapid transmission of a wave through it. For it results from the theory of sound that in media of different elasticities (so understood), but similarly constituted in other respects, these forces are to each other as *the squares* of the velocities with which the waves travel: so that the elastic force of the air would require to be increased in the proportion of the square of 901,000 (*i.e.*, 811,801 millions) to 1, to produce an equal velocity. Even this enormous number must be still further increased, since the velocity of sound is augmented by a peculiarity in the constitution of air which we should hardly be justified in attributing to the luminiferous ether, in virtue of which its elasticity is increased by heat given out in the act of its compression, and without which the velocity of sound would be only 916 feet per second instead of 1090. Thus the number above arrived at has to be further increased in the proportion of the square of 1090 to that of 916, which brings it to 1,148,000,000,000. Let us suppose now that an amount of our etherial medium equal *in quantity of matter* to that which is contained in a cubic inch of air (which *weighs* about one-third of a grain) were enclosed in a cube of an inch in the side. *The bursting power* of air so enclosed we know to be 15 lbs. on each side of the cube. That of the imprisoned ether then would be 15 times the above immense number (or upwards of 17 billions) of pounds. Do what we will—adopt what hypotheses we please—there is no escape, in dealing with the phænomena of light, from these gigantic numbers; or from the conception of *enor-*

mous physical force in perpetual exertion at every point, through all the immensity of space.

(66.) As this is the conclusion we are landed in—(for the evidence for the truth of the undulatory doctrine, or something equivalent to it, accumulating, as we shall see, in all quarters, and in the most unexpected manner receiving confirmation from facts utterly uncontemplated by its originator, obliges us to look on this result as something more than a scientific rhodomontade)—we shall endeavour to present it to the conception of our readers in a point of view which may enable them to realize it more distinctly. All who know the nature of a barometer are aware that the column of mercury 30 inches in height sustained in its tube, is the equivalent of the pressure of the aërial ocean which covers us, on its sectional area; and is just sufficient to counterbalance the pressure, on an equal area, of an atmosphere five miles in height of air *everywhere of the same density as at the surface of the earth.* This height (five miles) is what is termed in Barometry, "the height of a homogeneous atmosphere," and affords a measure of what may be called the intrinsic elasticity of the air, of an exceedingly convenient nature; and which is received as a kind of natural unit, in Meteorology and Pneumatics. Substituting now light for sound, and for air the luminiferous ether, we should have for the corresponding height of our homogeneous atmosphere (gravity being supposed uniform) five and a half billions of miles, or about one-third of the distance to the nearest fixed star! The measure thus afforded of intrinsic elastic power is of the same kind as that afforded of the intrinsic *tensile strength* of a wire or thread of any material by the statement of how much in length *of itself* it can bear without breaking. It frees us from the necessity of any mental reference to the actual weight or specific gravity of the material, which in this case is the more necessary, as, though we suppose the ethereal molecules to possess *inertia*, we cannot suppose them affected by the force of gravitation.

(67.) There is yet another theory of light which might be proposed, in which, still retaining the idea of an ethereal medium, its constitution should be conceived as an indefinite number of regularly arranged equidistant points (mathematical localities) *absolutely fixed and immovable in space*, upon which, as on central pivots, the molecules of the ether, supposed *polar* in their constitution, like little magnets (but each with *three pairs of poles*, at the extremities of three axes at right angles to each other), should be capable of oscillating freely, as a compass-needle on its centre, but *in all directions*. Any one who will be at the trouble of arranging half a dozen small magnetic bars on pivots in the *linear* arrangement of the annexed figure, will at once perceive how any

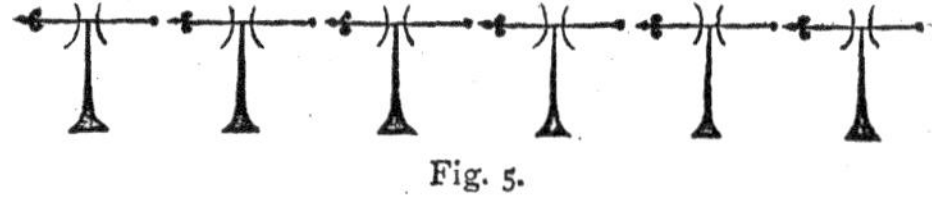

Fig. 5.

vibratory movement given to one, at any point of the chain, will run on, wave-fashion, both ways through its whole length. And he will not fail to notice that the bodily movement of each vibrating element will be *transverse to the direction of the propagated wave*—a condition which, as we shall hereafter see, is essential to be fulfilled in the luminous undulations. As this hypothesis, however, has hitherto received no discussion, and is here suggested only as one not unworthy of consideration, however strange its postulates, we shall not dwell on it; remarking only that every phænomenon of light points strongly to the conception of a solid rather than a fluid constitution of the luminiferous ether, in this sense,—*that none of its elementary molecules are to be supposed capable of interchanging places*, or of bodily transfer to any measurable distance from their own special and assigned localities in the universe. The constitution above suggested would merely superadd to this abstract idea of a solid structure, the further conception of polar forces bearing some general analogy to those which may possibly subsist among the gross particles of a tesseral crystal.

(68.) This would go to realize (in however unexpected a form), the ancient idea of a crystalline orb. And it deserves notice that under no conception but that of a solid can an *elastic and expansible* medium be *self-contained*.* If free to expand in all directions, it would require a bounding envelope of sufficient strength to resist its outward pressure. And to evade this by supposing it infinite in extent, is to solve a difficulty by words without ideas—to take refuge from it in the

* From a liquid the extreme particles would be constantly flying off in vapour and dissipating themselves in space.

simple negation of that which constitutes the difficulty. On the other hand, such a "crystalline orb" or "*firmament*" of solid matter conceived as a hollow shell of sufficient strength to sustain the internal *tension*, and filled with a medium *attractively, and not repulsively elastic*, might realize (without supposing a solid structure in the contained ether) the condition of transverse vibration; by establishing, *ipso facto*, lines of *tension* in every possible direction, along which undulations might be conveyed, like waves along a stretched cord, thus furnishing a fourth hypothesis, which, to those fond of such speculations may afford matter, *sui generis*, for consideration.

(69.) *Interference of the rays of light.* There is hardly a more beautiful or a more instructive object in nature than a large well-blown soap-bubble. Whether we consider the perfect regularity of its form, illustrating, as it does, in its exact equilibration the great mechanical laws to which the sun and planets owe their spherical figure—demonstrating, by its resistance to disruption by blasts of wind which distort it, and by its ready and complete resumption of its normal shape on their cessation, the powerful tensile force which holds it together; and proving, by the instantaneous collection of its filmy tissue into water-globules, in the act of bursting, the immense intrinsic energy of that force as compared with gravitation—to the mechanician it is fraught with matter of the highest interest. To the photologist, on the other hand, the vivid colours which glitter on its surface afford at once the simplest and most elegant optical illustration of the "law of interference" of the rays of light: a law we shall now proceed to explain, taking for our first exemplification of it this very phænomenon.

(70.) If a soap-bubble be blown in a clean circular saucer with a very smooth, even rim, well moistened with the soapy liquid,* and care be taken in the blowing that it be single, quite free from any small adhering bubbles, and somewhat more than hemispherical; so that, while it touches and springs from the rim all round, it shall somewhat overhang the saucer: and if in this state it be placed under a clear glass hemisphere or other transparent cover to defend it from gusts of air and prevent its drying too quickly;—the colours, which in the act of blowing wander irregularly over its surface, will be observed to arrange themselves into regular circles surrounding the highest point or vertex of the sphere. If the bubble be a *thick* one (*i.e.*, not blown to near the bursting point), only faint, or perhaps no colours at all will at first appear, but will gradually come on growing more full and vivid, and *that*, not by any par-

* M. Plateau gives the following recipé for such a liquid. 1. Dissolve one part, *by weight*, of *Marseilles soap*, cut into thin slices in forty parts of distilled water, and filter. Call the filtered liquid A. 2. Mix two parts, *by measure*, of pure glycerine with one part of the solution A, in a temperature of 66° Fahr., and after shaking them together long and violently, leave them at rest for some days. A clear liquid will settle, with a turbid one above. The lower is to be sucked out from beneath the upper with a siphon, taking the utmost care not to carry down any of the latter to mix with the clear fluid. A bubble blown with this will last several hours even in the open air. Or, the mixed liquid, after standing twenty-four hours, may be filtered.

ticular colour assuming a greater richness and depth of tint, but, by the gradual *withdrawal* of the faint tints from the vertex, while fresh, and more and more intense hues appear at that point, and open out into circular rings surrounding it; giving place as they enlarge to others still more brilliant, until at length a very bright white spot makes its appearance, quickly succeeded by a perfectly black one. Soon after the appearance of this the bubble bursts. During the whole process it has been growing gradually thinner by the slow descent of its liquid substance on all sides from the vertex, till at length the cohesion of the film at that point gives way under the general tension of the surface. The annular arrangement of the colours, and the coincidence of their common centre with this, the thinnest point of the film, evidently go to connect their tints with the thicknesses of that film at their points of manifestation, and to indicate that *a certain tint is developed at a certain thickness*, and at no other. This, we shall presently see, is really the case.

(71.) The order of the colours and the sequence of the tints is in all cases one and the same, provided the series be complete, *i.e.*, provided time has been given for the black central spot to form. Thus the *first* series, or order, contained within *the first ring* consists of black, very pale blue, brilliant white, very pale yellow, orange, red; *the second* of dark purple, blue, imperfect yellow-green, bright yellow, crimson; *the third* of purple, blue, grass green, fine yellow, pink, crimson; *the fourth* of bluish-green, pale pink inclining to yellow, red; *the fifth* pale bluish-green, white, pink. After these the colours grow paler and paler, alternately bluish-green and pink, and can hardly be traced beyond the seventh order.

(72.) None of these tints are *pure prismatic colours.* To see them to the best advantage the bubble with its glass shade should be placed out of direct sunshine, where only dispersed light, such as that of a cloudy sky, shall fall on it. Or, the illumination of the rings may be effected by a thin semi-transparent paper, or a ground-glass screen interposed between them and the incident light. And if, instead of illuminating *this* with the direct light of the sky, the coloured rays of a spectrum, formed by passing a sunbeam through a glass prism, be thrown upon it, the composite nature of their tints will be at once apparent. If all the rays but those at the red end of the spectrum be excluded from the illuminating beam, the rings will appear wholly red, separated by black intervals, and *much more numerous.* And if, now, the colour of the illuminating light be changed, so as to pass in succession through the whole prismatic scale of tints—orange, yellow, green, &c., from the red to the violet—the *colour* of the rings will undergo a corresponding change, the dividing intervals preserving their blackness, but their *number* still continuing greater than in white light. But, besides this, a very remarkable phænomenon will be observed. The rings *contract rapidly in diameter* as the colour of the illumination changes, being a maximum for a red and a minimum for a violet illumination; and if, by a slight movement given to the prism, the spectrum be made to traverse to and fro on the illuminating screen, the rings will appear to open and close in an exceedingly beautiful manner, undergoing at the same time a corresponding change of colour.

(73.) The composite nature of the rings, as seen in white light, is now abundantly clear. White light is a mixture of all the prismatic rays, and the set of rings seen in such light is of course a *mixture* of the several individual sets (*concentric, but differing by a regular gradation of size*), of all the several coloured elements of which white light consists. Imagine a painter who could "dip in the rainbow" and lay on, one after another, on the same paper and with the same centre, such a series of rings gradually decreasing in diameter, and each set tinted with the pure prismatic hue which corresponds to its size, from the extreme red to the extreme violet, in their proper degrees of intensity;—he would produce just such a series. If the diameters for all colours were alike, the compound rings would evidently be white and infinite in number, separated by black intervals. If they differed only a little—starting from a common origin—the first ring would be nearly white, but exhibiting a bluish border inwards and a reddish outwards, growing more and more "pronounced," and broken into intermediate tints in those beyond; but if considerable, the rings of different orders for different colours would soon mingle with and confuse each other's tints, creating the sensation of uniform whiteness: thus accounting for the comparative paucity of the mixed series.

(74.) In order, then, clearly to understand the nature of this phænomenon, it must be divested of this source of complexity, and studied in reference to light of one single colour or refrangibility—or, as it is called, "homogeneous" light, pure red or yellow, for instance. But before proceeding further, something more must be said of the whole class of phænomena referable to this head. And first, these colours are not dependent in any way on any colorific quality of the liquid of which the bubbles consist. Any sufficiently thin film, of any kind, suffices to produce them. They are seen in the oily scum on the surface of a stagnant pool. They are seen on the brilliant scales of old glass in stable windows, or on the wings of gaudy-coloured insects, or even on polished steel. Bubbles may be blown of a variety of liquids—nay, even of glass. However highly coloured, their intrinsic colour disappears when reduced to such extreme tenuity as is requisite for the purpose in question. But *all* exhibit *the same hues* in the same invariable order. Nay, more—it requires *no medium at all* to produce them, but only *an interval between two surfaces*. They are seen in the crack of a thick piece of glass which does not extend quite through its whole substance. They are seen when a piece of mica is partially split and one of the laminæ lifted up, following as a series of coloured lines the limit of the commencing fissure. It may be said that though no *solid or liquid medium* is here present, there is *air* between the divided surfaces. But under the exhausted receiver of an air-pump, there is no diminution of the colours, or alteration of their forms. It is to the *interval between the surfaces* that we have to look for their origin.

(75.) Here, then, we have LIGHT brought face to face with SPACE, and no escape! What happens at or between these surfaces? How is it that while a single surface reflects a dispersed beam of light indifferently over its whole extent, this indifference is destroyed by placing another reflecting surface behind it; and the reflexion (at least the *effective reflexion as regards the spectator*) rendered impossible when the second surface is at a certain distance, *or at certain distances*, from the first; while if placed at intermediate distances, it is either not at all affected, or only to a certain extent enfeebled? and *that*, when there is nothing, or at least nothing realizable to any of our methods of observation, between them? This is the problem before us, reduced to its simplest terms,—a problem which the corpuscular theory of light resolves imperfectly and unsatisfactorily, and the undulatory fully and without reserve.

(76.) When instead of using the prismatic spectrum (of which it is next to impossible to insulate from the rest a ray of perfectly definite refrangibility) to illuminate the film, we employ artificial light (such as that of a spirit lamp with a salted wick, which may be considered as almost perfectly homogeneous), the rings are seen with extraordinary sharpness; their central spot and their divisions having the blackness of ink, and absolutely innumerable; being traceable with a magnifier when too close to be otherwise distinguishable, apparently without limit. Thus disembarrassed of the complexity of overlapping rings of several colours, the phænomenon now is studied to greater advantage, and its explanation on either theory is more readily intelligible. That afforded by the corpuscular theory (supplemented by the Newtonian hypothesis of the fits of easy reflexion and transmission) is very simple and obvious. These "fits" or *phases*, it will be remembered, are supposed *periodically recurrent—i.e.*, succeed one another, or rather are repeated over and over again, in the same order and intensity, at equal intervals *of time*. The same *phase* then will recur to the corpuscules, at equidistant points *of space* in their progress through any uniform medium (in which the velocity of light is constant). Where the thickness of the film is *nil*, or so very minute as to bear no comparison to the distance which separates two of the equidistant points, it is obvious that having passed one surface they will still be in a state to pass through another, and will therefore *not* be reflected, so that in that case the reflected illumination of the first surface will receive no augmentation from light reflected at the second. The same is true if the thickness of the film be exactly that of two such equidistant points, or its double, triple, &c., for in those cases the corpuscule will arrive at the second surface in the same state and with the same dispositions as to reflexion or transmission as at the first; and therefore, having penetrated the first, will also penetrate the second. On the other hand, for thicknesses of the film exactly intermediate between these, the corpuscule on arrival at the second surface will be exactly in the opposite state or disposition. Having been transmitted then at the first, it will be reflected at the second, and, having in its passage back through *another equal* thickness reassumed its original state in which it first entered, it will there be *transmitted*, and so will reinforce by its light the general reflected illumination of that surface. Since the per-centage of the total light reflected at any transparent medium is but trifling, the light so sent back from the second surface will be nearly equal to that reflected from the first. Thus for these exactly intermediate thicknesses, the joint-reflected illumination is very nearly doubled, and between these and the former series of thicknesses will increase and diminish alternately and gradually.

(77.) Suppose now in the case of our soap-bubble the thickness of the film to increase uniformly outwards from its vertex (where it is nearly *nil*). Then it is evident that when exposed to dispersed light it will appear divided into equivalent circular zones alternately bright, and *comparatively* dark, the centre being also dark. And here we have a representation of our observed rings, with, however, this remarkable and most important difference, viz.: that the central spot and the dark divisions, on this explanation, ought not to appear absolutely black, but *half bright*, when compared with the brightest portions between them. In point of fact, some exceedingly slight reflexion *is* perceivable in the dark centre, but instead of half, it cannot be estimated at the fiftieth part of the illumination of the bright ring which immediately adjoins it.

(78.) As the thickness of a soap-bubble cannot be subjected to direct measurement, it is impracticable, in its case, to verify what must be considered the fundamental principle of this explanation —viz., the regular increase, *in arithmetical progression*, of the thicknesses at which the several successive black or bright rings appear. But of this we may satisfy ourselves, by adopting a different mode of producing and viewing them. When a spectacle-glass or any other convex glass *lens* of long focus is laid

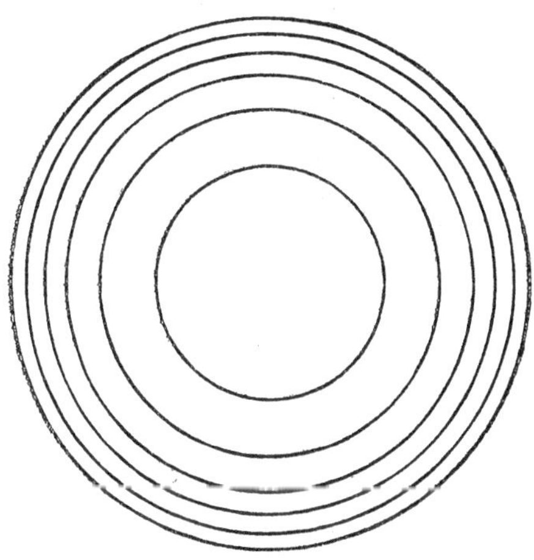

Fig. 6.

down upon a plane glass before an open window (both being scrupulously clean and well polished), and a slight pressure applied, the same dark spot, surrounded by the same series of coloured rings, is seen, their centre being at the point of contact of the glasses, where of course their distance is *nil*. If the focal length of the lens be not *very* large, they will require a magnifier to be well seen, their diameters being in that case very small; but with a lens of 20 or 30 feet focal length it is considerable, and the rings may be seen, and their diameters measured, with ease. Now it is found that these *diameters*, for the first, second, third, &c., dark rings in order (reckoned from the centre), are not in the proportion of the numbers 1, 2, 3, &c., but of the numbers 1, 1·414, 1·732, 2·000, &c., which are their exact *square roots*, giving to their system the appearance represented in the preceding diagram; and this is exactly the progression of distances from the point of contact measured on the surface of the plane glass which correspond to the series of *perpendicular distances between it and the convex spherical surface of the upper glass* in the proportion of the arithmetical series, as may be seen in Fig. 7.

(79.) So far, then, the Newtonian hypothesis affords a satisfactory account of the facts; in all, that is, but that one particular already adverted to. This, however, must be considered as conclusive against it; while, on a consideration of the whole case, there remains outstanding this strange *fact*—that at certain distances between two partially reflecting surfaces, forming a regular arithmetical progression from *nil* upwards, the portion of a beam of light reflected from the second, after passing back through the first, so far from augmenting the first reflected light, *annihilates it*, and furnishes us with an instance (which is, as we shall see hereafter, not the only one) of *the combination of lights creating darkness!*

(80.) The question now arises,—Will the undulatory

theory help us in this difficulty, while at the same time rendering an equally satisfactory account of the other facts? To this we are enabled to reply in the affirmative. Two equal *sounds* we know, under certain circumstances, *can* produce silence, as when the two strings which, in a pianoforte, go to produce, when exactly in unison, a uniform and liquid note; if very slightly out of tune, produce what are called *beats*, or a succession more or less rapid (accordingly as the strings are more or less discordant) of sound and silence. The same tide-wave arriving at the same spots in the sea by two courses of different lengths, results in producing *no rise and fall of the water at all*, if the difference of path be such that the high water of one portion shall reach

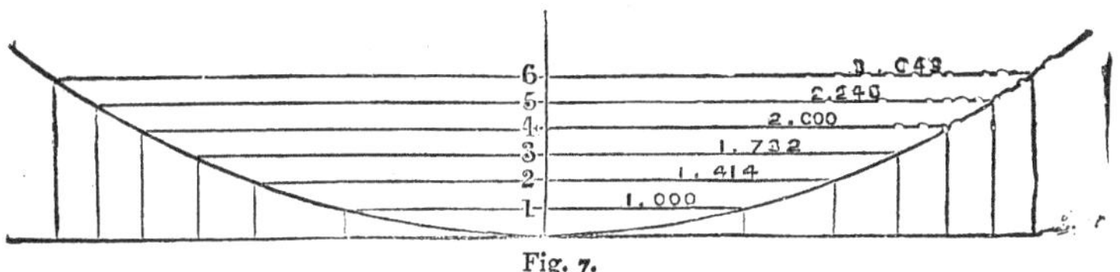

Fig. 7.

the place at the same moment with the low water of the other. This is the case at a point in the North Sea, midway between Lowestoft and the coast of Holland, in lat. 52° 27′ N., long. 3° 14′ E. Its position was pointed out by Dr Whewell from theory, and the fact verified by Captain Hewett, R.N.

(81.) This latter exemplification contains the essential principle of the explanation in question, in nearly its simplest state. If two waves, or rather two regular series of equal waves all exactly like one another, *and all having set out initially from one common origin*, reach the same point by two different channels or lines of com-

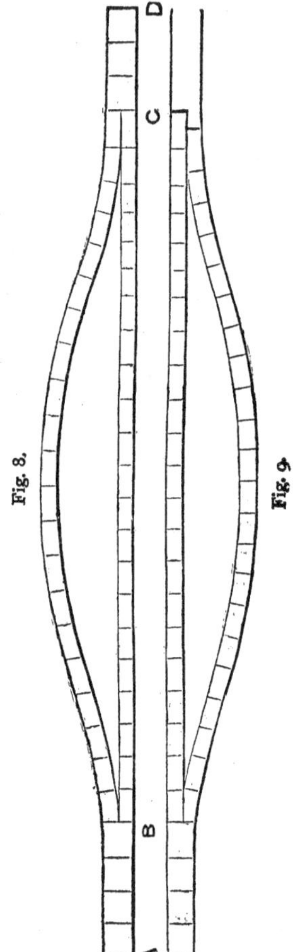

Fig. 8.
Fig. 9.

munication differing just so much in length that the

crests of the one series shall reach it at the same identical moment with the *crests* of the other; the two series of crests conspiring and being superposed on each other will produce crests of double the height of either singly: while, on the other hand, if the difference of channels be such that the crests of the first series shall reach it simultaneously with the troughs (or lowest depressions) of the second; the one will destroy the other, and there will be neither elevation nor depression at their joint point of arrival. In the former case, supposing the two channels thenceforward to unite into one (as in the annexed figures, which require no explanation further than that the series of cross lines represent the *crests* of the waves), the two series when they reunite in a channel C D, as in Fig. 8, the exact size of the initial one, A B, will form a joint series exactly similar to that in A B, which will run on in that channel thenceforward; but in the latter, as in Fig. 9, there will be produced no waves at all, and the water in C D will (except just close to the point of junction, where some kind of eddy will be formed) remain undisturbed.

(82.) Accepting the term "wave" in its most general sense, in whatever way we suppose it propagated, whether by alternate up-and-down movements of the successive particles, as in water-waves—by transverse lateral ones, as in a stretched cord *wagged* horizontally—or by direct to-and-fro vibration, as in the air-waves, in which sound consists—or in any more complex manner, the same considerations evidently apply. If two sets of exactly equal and similar waves can by any previous arrangement be made to arrive *simultaneously* at the "entrance" of one and the same "channel" (using these terms also in their most general sense) along which each set, separately, might be freely propagated—*i.e.*, so that the foremost crest of the first set shall strike the mouth of the channel at the same moment with that of the other—they will combine and run on along the channel as a single set or series of waves of double the height or intensity. In this case they are said to arrive "*in the same phase*" (a term borrowed from the phases of the moon which passes periodically through the states of full and new, increase and wane). The same will be the case if the foremost *crest* of the series be so timed (by the previous arrangements) as to reach the mouth of the channel simultaneously with the second, third, or fourth *crest* of the other, in which case the one set is said to be in arrear or advance of the other (as the case may be) by one, two, or more entire "*undulations.*" On the other hand, if the foremost *crest* of the one set be so timed as to arrive simultaneously with the first, second, third, &c., *trough* of the other—up to the time of its arrival indeed the one, two, or three foremost waves which are not contradicted will run forward;—but, from the moment when the others begin to arrive, they will cease to be followed up by any more. In this case the one set is in advance or arrear of the other by exactly one, or three, or five, &c., *semi-undulations*, and the series are said to be *in opposite phases.* In the intermediate "phases" it is easy to see that a combined set of waves *will* be produced, but intermediate in height, intensity, or (as it is called) "amplitude" between these two extremes

—viz., *nil* on the one hand, and *reduplication* on the other.

(83.) The vibrations by which light and *musical* sounds (*to which light is analogous*) are conveyed are so exceedingly minute, and the shock conveyed by each separately to our nerves, in consequence, so small, that it requires a continued series of them to impress our senses. The first few vibrations therefore which run on "*uninterfered* with" produce no sensation, and are as if they existed not. And thus we see how it may happen, that in the case of a complete *opposition of phase* two equal musical sounds may produce silence, and two equal rays of light *complete and continued darkness;* that a perfect coincidence of phase has the effect of doubling the sensation; and the intermediate states, a greater or less intensity as the case may be, short of that limit.

(84.) Let us now proceed to apply our principle (that of "*superposed and* INTERFERING VIBRATIONS") to the matter in hand. Suppose a series of equal and equidistant light-waves (such as *a ray* of homogeneous light is in this theory always understood to mean) to fall perpendicularly upon a plate of any transparent medium. A certain very small per-centage of it will be reflected back by the first surface—that is to say, a series of similar undulations, but of much less intensity or "amplitude," will be propagated back from the point of incidence. The remainder of the total movement thus subdivided will pass on, and, arriving at the second surface, again a *very nearly equal* series (the per-centage being the same, and the total incident light having suffered very little diminution) will be returned, and, passing back through the first surface (with only the same trifling *per-centage* of loss), will, on emerging, *be superposed on the first reflected series*, with which it will be coincident in direction. If then the thickness of the plate be such that in passing to the hinder surface, back again, and out at the first, the second series shall have *lost upon the first* precisely one, three, five, or any odd number of *semi*-undulations, it will begin to emerge in the exact opposite "phase" of its period to that of the undulation of the first reflected series which starts from the same point at the same instant. Here, then, we have the case contemplated above of two series of equal waves entering the same "channel" in opposite phases. They will therefore destroy each other, or the intensity of the joint ray will be *nil*. The contrary will happen if the thickness be such as to produce a retardation of two, four, or any even number of semi-undulations. In that case the two reflected rays will conspire and produce a joint one of double intensity: and of intermediate in the various cases of intermediate retardation.

(85.) Thus we see that the degree of brightness of the reflected light depends on the thickness of the reflecting film, and that for a certain series of thicknesses *in arithmetical progression*, the joint reflection is *nil;* for another series exactly intermediate, it attains a maximum of intensity; and between these limits, all gradations of illumination will arise according to the intermediate thicknesses supposed to exist. This is so far in general accordance with the phænomena described: but before applying it to the case of our coloured rings, it must be mentioned that in reckoning the number of undulations or semi-undulations by which the second reflected ray actually is in arrear of the first on emergence, we have to consider the different modes in which the reflexion of a wave is accomplished at the surface of a medium denser or rarer than that in which it moves and is reflected. To present this clearly, we will take the most familiar illustration—that of the propagation of motion by the collision of elastic balls. Imagine a great number of equal ivory balls (supposed perfectly elastic) *in contact*, but *connected* only by an elastic string passing centrally through each and along the common axis of all; and pinned or fastened to each *at its centre* so that the separation of any two shall stretch only that part of the string between their centres. Suppose now that a shock is given to the extreme ball at one end in the direction of the common axis, by another similar and equal detached ball driven against it. By the received laws of elastic collision it will give up *its whole motion* to that which it first strikes, and be itself reduced to rest. In like manner the motion so communicated to the first will be handed on undiminished to the second; itself resting *and therefore* remaining in contact with the striking ball, and so on. Thus what may be termed a "wave of compression," will run along the series till it reaches the ball at the other end. This, having none in front to communicate its motion to, will start off; and, were it free, would quit the series. But this it cannot do, by reason of the elastic thread; which however it will stretch in its effort to do so, and be ultimately brought back by its pull. But in so doing the same pull will also be communicated to the ball behind it, drawing it forward, and so in succession to those yet behind; and in this manner, a *wave of extension* will run back along the series. If the tension of the string be very violent (suppose equal to the repulsive elasticity of the balls), this wave will run back with *the same velocity* as the other. Here we have then a case of the reflexion of a wave, where, *in the very instant* of reflexion, its character is changed *ipso facto* from that of a wave of compression to one of extension—in other words, it starts backwards in *the opposite phase* to that of its arrival; or, again in other words, a *semi-undulation is lost or gained* (for it matters not which) *in the act of reflexion.*

(86.) This is the extreme case of reflexion from a denser medium on a rarer—for here there is absolutely nothing to carry on the motion beyond the terminal ball. Such a case never occurs in nature as regards light; since even what we call a vacuum is filled with the luminiferous ether. To assimilate it to such as *do* occur, suppose a second series of *smaller* balls, similarly connected with each other, *but not with the first set*, and brought end to end with it, with just room between for one intermediate free ball of the smaller size to play backwards and forwards as a go-between; and let this, in the first instance, be placed in contact with the last ball of the first set. When the movement reaches it, it will be driven off, and immediately striking the end ball of the second set, will propagate along *it* a wave of compression, coming itself to rest. In so doing it will carry off *some, but not all* of the motion of the terminal ball of the first set. This will still continue to advance after the blow, but to a less extent, and with less momentum than in the former case, and, just as in that case, will propagate backward a wave, though a feebler one, *of extension.* Starting, then, from the same place at the same moment, the two waves—the reflected portion (or echo) and that which runs forward in the second set of balls, set out each in its own direction *in opposite phases.*

(87.) The intensity of the reflected wave or echo will be feebler the nearer the balls of the two sets approach to equality (or the less the difference of density in the two media). If they are exactly equal, the go-between ball will carry off all the motion of the ball which strikes it—or there will be *no* reflected wave, no echo. And this agrees with fact. At the common surface of two transparent media of equal refractive power, however they may differ in other respects, there is no reflexion. But suppose the second set of balls, as also the single intermediate one, *larger* than the first. In that case (still according to the laws of elastic collision) the last ball of the first set not only will *not advance* after the shock, but will be driven back, and the wave which it will propagate backwards will no longer be one of extension, but *of compression.* This being also the case with that propagated onwards in the second series,—in this case both will start on their respective courses from the point of reflexion *in the same phase.*

(88.) In the undulatory theory of light the "denser"

medium corresponds to the series of larger balls in this illustration. This ought to be so, for the velocity is less in the denser medium as it is in the larger of two balls after their collision; and because, as already remarked, the ether in such media must be either denser in proportion to its elastic force, or somehow encumbered by their material atoms. And hence we finally conclude that in the act of the reflexion of light on the surface of a rarer medium, the phase of the undulation changes, and a *semi-undulation* is lost (or gained—it matters not which): but not so when the light is reflected from a denser medium.

(89.) To return now to the case of a thin pellucid film. If its thickness, *i.e.*, the interval separating its two surfaces, be any number of *semi*-undulations; double that number, *i.e.*, an exact number of entire waves, will have been lost by the wave reflected from the second surface at its re-emergence from the first, by reason of its greater length of path; and thus were no part of an undulation lost or gained *in the act of reflexion*, it would start thence in exact harmony with the first reflected ray. But the second reflexion being made at the surface of a rarer medium, an additional *semi-undulation* will have been lost, so that the two reflected rays will really start from the first surface in *complete discordance*, and destroy each other. The same is the case if the thickness be *nil*, or so excessively minute as to be much less than the length of a wave, as at the vertex of a soap-bubble when just about to burst. Here also will the same mutual destruction of the reflected waves take place. And thus we have explained the complete or very nearly complete darkness of the central spot, and of a series of rings corresponding to thickness of 1, 2, 3, or more semi-wave lengths. At the intermediate thicknesses (*i.e.*, of 1, 3, 5, &c., *quarter*-wave-lengths) the exact reverse will happen—the reflected rays will start together in harmony and appear as a ray of double intensity, thus explaining the intermediate bright rings.

(90.) In the case of the rings produced between two glasses of the same material, the intermediate film being *air*, it is the reflexion from its *first* surface, not its second, that is effected from a rarer medium; so that it is at this surface that the additional *semi*-undulation is *gained* by the first reflected ray. In all other respects the reasoning is the same in both cases, and the explanation equally complete in both.

(91.) It will be perceived that we have not been sparing of words in this explanation. The epigrammatic style is ill-suited to clearness in the exposition of a principle which it is essential to seize with perfect distinctness, and in seizing which considerable difficulty is commonly experienced. If any doubt or misgiving, however, should still linger on the mind as to the *applicability* of the analogy by which the loss of half an undulation necessitated by the blackness of the central spot has been explained, a simple but striking experiment will suffice to dissipate it. Let a set of rings be formed by interposing, between two glasses of very different refractive densities, a film of liquid intermediate in that respect—as, for instance, oil of sassafras between a lens of light crown, and a plate of heavy flint-glass. In this case the reflexions from the two surfaces are performed either *both* from a denser medium upon a rarer, or *both* from a rarer on a denser according as one or the other glass is uppermost. In the former case *two* semi-(or one entire) undulation will be gained or lost between the reflected rays at emergence, in addition to the entire ones lost *between* the glasses: in the latter none. At the central spot, then, the two reflected rays will start on their backward course in exact harmony, and the spot will be white, not black; and a similar reversal of character will of course pervade the whole series of rings. This result, predicted by theory, has been found confirmed by experiment.

(92.) It was a favourite idea of Newton that the colours of all natural bodies are in fact the colours of thin pellucid particles of such sizes and thicknesses as to reflect those tints which, in the scale of tints of the coloured rings above described, most nearly correspond to them. This idea we know now to be untenable, if for no other reason than that we are sure the ultimate particles or indivisible atoms of bodies (if any such there be), are at all events many hundreds, thousands, or millions of times smaller than even a single wave-length of any homogeneous ray of light. It will, of course, be asked how we know these wave-lengths. And this we must now explain: in doing which we shall have to develop the most astounding facts in the way of numerical statement which physical science has yet revealed.

(93.) In a series of equal waves running along still

water, the "wave-length," or "length of an entire undulation," is the linear distance between two consecutive *crests* or two consecutive *troughs*. This is its simplest conception, and it will suffice for our immediate purpose. The waves being equal and similar will all run on with the same velocity, which may be ascertained by noticing how long any one takes to run over a measured distance on the surface, or the distance run over in a determinable time, suppose a second And if at the same time we note the number of waves whose crests pass a fixed point (a float, for instance) in the surface, in a second of time the interval between two consecutive crests will of course become known. And *vice versâ*, if this interval be known, and the velocity of the waves; the number of undulations passing the float per second is easily calculated. Now *this number is necessarily identical with that of the periodically reciprocating movements or vibrations of the first mover* (whatever it be) *by which the waves are originally excited.* This continuing the same, the same number of waves will pass the float in the same time, whatever be their velocity of propagation. Of these three things—the velocity of propagation, the number of alternating movements, waves, or pulses per second, and the linear interval between two consecutive ones—any two being given, the third is easily calculated. For example, a string sounding a certain note C in the musical scale makes 256 *complete* oscillations to and fro, per second. As each of these sends forward an air-wave consisting of a semi-wave of *compression* by which the particles of air advance, and another of *expansion* by which they return to their places, these waves, 256 in number, being all comprised in, and exactly filling the distance (1090 feet) run over by sound in that time, each entire wave will occupy 4·254 ft. And, *vice versâ*, if we knew this *à priori* to be the wave-length, we should rightly conclude 256 to be the number of complete vibrations or pulses per second.

(94.) Mechanical processes enable us to grind and polish a glass surface into the segment of a sphere of any required radius—as well as to a plane almost mathematically true. Suppose such a glass surface worked, we will say, to a sphere of 100 feet radius, to be laid (convexity downwards) on a truly plane glass. The coloured rings will be formed, as above described, about a central dark spot; and if illuminated, instead of ordinary daylight, by the prismatic rays, in succession, a series of simply bright and dark rings of the several colours in their order will be formed, whose diameters in different series will correspond to their respective tints. Under these circumstances, the linear measurement of these diameters may be performed with ease and with great precision. Now these diameters are the *chords* of arcs of a circle on a radius of 100 feet represented in fig. 7, by the horizontal lines, the *versed sines of whose halves* corresponding (represented by the perpendicular lines) are the distances between the glasses at those points, or the thicknesses of the interposed film *of air*, and are easily calculated when the radius and the chords are known. On executing the measurements it is found that these distances, reckoning outwards and commencing with the centre, do actually follow the law of arithmetical progression (as on the above theory they should do), being in the proportions of the numbers 0, 1, 2, 3, etc.

(95.) By measuring then the diameter of (say) the tenth dark ring (for the sake of greater precision), calculating the corresponding interval, *or versed sine*, and taking one-tenth of the result, we shall get the interval corresponding to the first dark ring—for any particular coloured light—and this, by what has been above shown, is the half of a *wave-length* for such light. Proceeding thus, Newton found for what he considered the most luminous yellow rays, one 89,000th part of an inch for the interval in question, which gives for the length of an entire undulation of such rays, one 44,500th of an inch. This comes exceedingly near to the result which later experimenters have obtained for that purely homogeneous yellow light emitted by a salted spirit-lamp, which is one 43,197th of an inch. For the extreme red and extreme violet rays, (as well as their limits can be fixed,) the corresponding wave-lengths are respectively one 33,866th, and one 70,555th of an inch.

(96.) These, it will be observed, are the lengths of the undulations in air. In water, glass, or other media, they are smaller, in the inverse proportion of the refractive index of the medium; for in such media the velocity of light, as we have seen, is less in that proportion; and the number of undulations per second remaining the same, while the space occupied by them is less, their individual extent must of course be less in the same proportion. This, too, is in accordance with experiment. If water, oil, or any other liquid be introduced between the glasses, the rings are observed to shrink in diameter, and the more so (and to the exact extent required by theory) the greater the refractive power of the liquid.

(97.) If the sensation of colour be, in analogy to that of tone or musical pitch, dependent on the frequency of the vibrational movements conveyed to our nerves of sensation, it becomes highly interesting to ascertain their *degree of frequency*, in order to establish the relation between the two senses of hearing and seeing in that respect. The ear, we know, can discriminate tones only between certain limits, comprising about nine octaves, the lowest sound audible as a note making about 16, and the highest about 8200 vibrations per second. Taking the velocity of light (as above) at 186,000 miles* per second, and reckoning 33,866 wave-breadths to the inch for the extreme red, 43,197 for the soda-yellow, and 70,555 for the extreme violet, we find for the impulses on the retina per second which produce these sensations of colour, respectively, the following enormous numbers:—

Extreme red, . . .	399,101,000,000,000
Soda-yellow, . . .	509,069,000,000,000
Extreme violet, . . .	831,479,000,000,000

These extremes are nearly in the proportion of 2 to 1, so that the whole range of visual sensation on this view of the subject is comprised in about one octave. If the

* Roughly, 1000 million feet.

ear could appretiate vibrations of this degree of frequency, the sensation corresponding to the middle ray of the spectrum would be that of a note about forty-two octaves above the middle C of a pianoforte.

(98.) In each of these inconceivably minute intervals of time (compared to which a single second is a sort of eternity), *a process has been gone through by every molecule of the ether concerned in the propagation of the ray:* a process as strictly definite, as exactly regulated, as the movement of a drop of the ocean in its conveyance of the tide-wave. Taking up its motion from the particle immediately behind it (whose movement it exactly imitates), and transmitting it on to that immediately before, it starts from rest, not suddenly, with a jerk, but (under the strict control of those elastic forces already mentioned), increasing gradually in speed to a maximum—then, as gradually, relaxing, coming to a momentary rest, and retreating to its original position by the same series of measured gradations in reverse order, to be ready in its place for the reception of the next impulse. Nor does it seem possible to avoid the conclusion, if we trace up the movement to its commencement—to the source of the light—the *material particle* in whose combustion or incandescence it originates—that such is the actual vibratory movement of that particle itself. And thus we are brought into the presence of the working of that mechanism by which flame and incandescence ("φλογωπον σημα πυρος—the brilliant *miracle* of fire," as the Greek poet* not inaptly terms it) are pro-

* Æschylus.

duced. In the disruption of one chemical combination, and the constitution of another, a movement of mutual approach, more or less direct, is communicated to the uniting molecules, which, under the influence of enormous coercive powers, is converted into a series of tremulous, vibratory, or circulating movements communicated from them to the luminiferous ether, and so dispersed through space. Incandescence without combustion (as in a piece of red-hot iron) must be looked upon, from this point of view, as a result of the continuance of this vibratory movement after the primary exciting cause has ceased, and of its gradual decay by communication to the surrounding ether; as a musical string continues to sound after the blow which set it in motion, till gradually brought to rest by the surrounding air.

(99.) This may perhaps appear a digression from our subject. But it will be recollected that our object in these lectures is not to produce a treatise on optics, but to fix attention on the immensity of the forces in action, and the minuteness and delicacy of the mechanism which they animate in the most ordinary operations of Nature, and which the phænomena of light have been the means of revealing to us. We have no means, indeed, of measuring the *actual* intensity of the "coercive forces" so called into action in the excitement of a luminous vibration, but that we are fully justified in applying to them the epithet "enormous," the following consideration will suffice to show. Whatever be the extreme distance of excursion to which a vibrating molecule is carried from its point of repose, or its medium situation, in the act of vibration; the acting or coercive force must suffice to bring it back from that distance in one fourth part of that inconceivably minute fraction of a second by which, as above shown, the period of a complete vibration is expressed. Taking the case, then, of any particular ray (as for instance that between the green and blue rays of the spectrum, corresponding to a wave-length of one 50,000th of an inch, and to a period of one 589 billionth of a second), if we assume the extent of excursion, we can very readily calculate the intensity of the force (as compared with that of gravitation) which, acting uniformly during that time, would urge it through that space. Let us suppose then, that the nerves of the retina are so constituted as to be sensibly affected by a vibratory movement of no greater extent or *amplitude* than one trillionth* part of an inch either way; and the calculation executed, we shall find that a force exceeding that of gravity in the proportion of nearly thirty thousand millions to one must be called into action to keep up such a movement. Our choice lies between two immensities, we had almost said between two infinities. If we would bring the force within the limits of human comprehension, we must in the same proportion exaggerate the delicacy of our nervous mechanism, and *vice versâ*.†

* A trillion is a million of billions = 10^{18}, or 1,000,000,000,000,000,000.

† The hypothesis of a uniform action of the coercive force in the text is only assumed for the convenience of such of my readers

(100.) Connected with the colours of thin plates are several distinct classes of optical phænomena, in which colours of the same kind, and explicable on the same principle, arise in the reflexion and transmission of light through or between pellucid plates of considerable thickness, or through spherical drops of water, examples of which are to be observed in the pink and green fringes which are often seen bordering the interior of a rainbow, and in those similarly coloured fringes (of exceedingly rare occurrence) which sometimes run, like a bordering ribband, just within the contour of a thin white cloud in the near neighbourhood of the sun. Upon this class of phænomena, however, we shall not dwell further than to observe that they prove the law of the periodical recurrence of similar phases at equal intervals, not to be confined to very minute distances in the immediate neighbourhood of reflecting or refracting surfaces, but to extend over the whole course of a ray of light—as, on the undulatory theory, it necessarily must do. We now,

whose knowledge of dynamics is limited by this very elementary application. Properly speaking, we ought to assume the coercive force to vary in the direct ratio of the distance; on which supposition only will large and small vibrations be executed in equal times. Calculating on this (the correct) principle, and taking the extreme excursion (as in the text) at one-trillionth of an inch, the ratio of the coercive force to gravity at that distance will be found as 35.465,000,000 to 1. On the other hand, as a strange contrast to the immensity of such a force, we shall find the maximum velocity it will have generated on the arrival of the molecule at the medial point of its vibration not to exceed 1-270th part of an inch per second!

therefore, proceed to the next branch of our general subject, that of the Diffraction of Light.

(101.) DIFFRACTION.—The optical phænomena which refer themselves to this head are many and various. They are not, for the most part, very obvious, but are exceedingly curious and interesting in their details, and some of them, under careful arrangement and with good optical appliances, very brilliant. Familiar examples offer themselves in the twinkling of the stars and the changes of colour they exhibit during the different phases of their scintillations:—in the vivid radiating streaks of light which seem to stream outwards from any small and *dazzlingly* brilliant point of light (as for instance the reflection of the sun on a small polished globe, as a thermometer ball):—in the colours exhibited when a bright point is seen reflected on or refracted through a surface regularly striated or scratched across with fine equidistant lines, as beautifully exhibited in the so-called "Barton's buttons" (from the name of the ingenious and skilful amateur mechanist who first executed them); brass or steel buttons delicately cross-lined by engine work:—in the lateral images of a candle seen reflected on polished mother-of pearl: and in the coloured halos often seen to surround the flame of a candle in certain states of the eye and their artificial imitations in a mode presently to be described. Less obvious to common observation, and requiring particular arrangement, instrumental or otherwise, to see them distinctly, are the phænomena (referable to the head of diffraction) of the rings and other appendages seen to surround the images of stars when viewed through telescopes of great magnifying power, and with apertures either of the usual circular form, or of forms otherwise varied for this express purpose. And though last in our order of enumeration, by far the most important and instructive as elucidatory of the principle of explanation, applicable to all the phænomena of this class;—the coloured fringes seen to follow the outlines of shadows when thrown by a light emanating from an extremely small but intensely luminous point. With these, therefore, we shall begin.

(102.) It was objected to the undulatory theory of light by Newton himself, that sound, to which that theory assimilates it, spreads from an aperture through which it is transmitted, or round the edge of an intercepting screen of any kind, equally in all directions; and thus, were the analogy exact, there could be no shadows. The objection is founded partly on an imperfect statement of the fact, and partly on omitting to allow for possible differences in the natures of the conveying media, and in the modes of vibratory motion conveyed. Every one is familiar with the sudden outbreak of sound from a railway train heard at a great distance when it emerges from a cutting, or turns the corner of a wall or of a hill. Sound is propagated through water with greater sharpness, velocity, and distinctness, than through air. But an obstacle interposed under water, as a projecting pier, or a rock, cuts off the *rays of sound*, as appears from direct experiment, with much greater definiteness than in air, and casts, so to speak, an evident acoustic shadow. Nor will it appear at all surprising that an effect of this kind should, in the case of light, be carried still farther when we consider that the aërial impulses by which sound is propagated, take place *in the direction* of the sound-ray, so that in passing (for instance) through an aperture in a screen, a quantity of air is *pushed bodily through it*, and issuing on the other side, causes an increase of *local* density due to the actual introduction of additional air at a given spot, which of course tends to *expand laterally* as well as to *push forward*, and is not restrained from so doing by the lateral pressure of the rest of the wave, which is suppressed. Light, as we have already intimated, is propagated through an elastic medium more in analogy with a solid than a fluid, (which Newton's objection implies,) and by vibrational movements not in the direction of the ray, but transverse to it, so that in its passage through an aperture, or beside the edge of an obstacle, *this* cause of lateral spreading, at least, is absent; whatever other this peculiar mode of propagation may call into action. Lastly, however, the phænomena of diffraction with which we are now concerned rely for their explanation on this very principle—that shadows are *not* strictly definite, and that there really *is* a certain, and not very small amount of lateral spreading of the light into the space occupied by what may be called the geometrical shadow.

(103.) If a room be darkened and the sun allowed to shine into it only through a very small aperture, as a pin-hole, the rays which emanate from different points of its apparent disc, passing straight through and crossing at the hole, will depict on a white screen held at a distance of several feet from the hole a circular *image* or inverted picture of the sun, which may be considered as the circular base of a cone of rays having the pin-hole for its vertex. In this case, the illumination of the screen, if placed at a great distance, is feeble, and, if near, the circular patch of light inconveniently small. But if instead of a pin-hole, be substituted a convex glass lens of short focus, the whole of the sun-light received on it will be concentrated in the very small image of the sun formed in its focus, and, diverging *thence* will spread out into a much wider cone of light, and form a much larger circular and brilliantly illuminated area on the screen, affording every facility for the examination of the shadows of objects thrown upon it; with the additional convenience that by a reflector outside of the window, the illuminating sunbeam may be thrown horizontally, at whatever time of day the experiment is made.

(104.) The condition essential to the distinct exhibition of all phænomena of this class—that of a very brilliant light emanating from a very small point—being thus secured, let an opake body of any form be placed between the point and the screen, so as to cast a shadow on it. It would naturally be expected under such circumstances that the termination of the shadow on all sides should be a clear and sharply-marked outline, separating a uniformly bright space on the outside from a uniformly dark one within, and free from that external gradation from light to darkness which constitutes what is called the *penumbra* in ordinary shadows, which arises from the angular diameter of the sun.* Quite otherwise. A shadow indeed *is* formed, but instead of a sharp and sudden transition from darkness to light, it terminates in three coloured fringes, following its contour, the inner being the broadest and more distinctly coloured, the outer extremely faint and feebly tinted. The order of the colours, reckoning from the first *dark* fringe, is, generally speaking, analogous to that of the colours of thin plates proceeding outwards from the dark centre rings, only degrading more rapidly, viz., blue within, and yellow and red without. And that the tints originate in the same way from the superposition of a series of dark and bright fringes of the different prismatic colours, of different breadths, is shown (as in the colours of thin plates) by throwing on the lens in succession the several coloured prismatic rays, when the fringes are seen in each colour much more numerously and sharply defined, being broadest in red light and narrowest in violet.

(105.) If the object casting the shadow be long and very narrow, as a hair or a strip of card not more than a 30th of an inch broad, the phænomena are still more curious and complex. Besides the exterior coloured fringes already described, others are seen *within the shadow*, running parallel to its length, similarly disposed along both its edges, and blending in the middle into a

* The diffracted fringes may be seen very well on the borders of shadows cast by the light of Venus when at its greatest brightness, on a white surface, in a room with a single window, and under favourable circumstances as to twilight.

central line devoid of colour. That these fringes originate in the mutual interference of rays which have passed beside *both* the edges of the object and entered the shadow, is proved by intercepting the light on one side only, leaving that on the other to pass freely. *All* the interior fringes and the central streak disappear, leaving only the *exterior ones on the illuminated side* outstanding. The shadow (which is now formed under the same circumstances as in the former case) must, it is clear, be still receiving one-half the total quantity of light which it did before; and if its edge be narrowly examined, it will be seen not to terminate in any sharply-defined line cutting it off from the fringes, but to graduate off insensibly; and hence arises a very singular phænomenon. If the light be readmitted on *both* sides, and the breadth of the shadow, or what under such circumstances must be accepted as such, be measured, it is found to be much broader than it ought to be were it limited by straight lines drawn from the illuminating point through the edges of the object. And, what is still more remarkable, if its breadth be measured on a screen, successively placed at different distances from the object, its increase of breadth is found not to be in the simple proportion of its increased distance; as it would were the aerial shadow (or the space shaded by the object) bounded by straight lines; but as if by curves starting from its edges, and having their convexities towards the light. And, finally, if the object and the screen, *preserving the same distance between them*, be moved gradually nearer and nearer to the illuminating point, the fringes, both interior and exterior, are observed to dilate in breadth, according to a certain law which it is not necessary here to state, but whose agreement with the result of calculation affords a very satisfactory verification of the theory adopted for the explanation of the whole series of these phænomena on the undulatory hypothesis: while, on the other hand, it is very obvious that no such dilatation could possibly take place were the fringes produced by any kind of action on the rays in their passage by or near to the edges of the object, in the nature of attractive or repulsive forces originating in the material substance of which it consists, and deflecting them from their rectilinear paths; inasmuch as such action *could* neither be increased nor diminished, or in any way modified, by the greater or less distance traversed by the rays *before* their arrival within its sphere.

(106.) The appearances exhibited when the light is *transmitted* through a narrow rectangular slit, are even more curious. In this case a bright image of the opening is thrown on the screen, but instead of being an evenly illuminated narrow band, bounded on either side by uniform darkness, it is bordered both externally and internally by parallel fringes. The external ones are bright and highly coloured, and vary in breadth, but not materially in brightness, as the screen is withdrawn from the slit; but the interior fringes undergo singular changes as the distance increases. At near distances they are narrow and close, and leave a medial space of uniform light; but as the distance increases they enlarge in breadth, and close in on the illuminated space, so that at a certain distance a medial *dark* line makes its appearance—which, if the distance be still further increased, changes to a bright one, and so on alternately till after a certain distance is attained the alternations cease. If the screen be stopped at the last position in which the medial line is dark, and there fixed, and an opake strip, exactly half the breadth of the slit, be held *medially* along its whole length, so as to divide the slit and reduce it to two parallel ones, each one quarter of the original breadth (by which, of course, the total light traversing the aperture will be reduced to half its amount),—instead of darkening still more the medial dark fringe on the screen, as would naturally be expected on throwing a shadow upon it, the very reverse happens: the dark fringe in question disappears altogether, and is replaced by a bright one.

(107.) If the shape of the body which casts the shadow be angular, having salient and re-entering angles; the fringes where they surround the re-entering angles cross and pursue their courses up to the shadows of the sides respectively opposite to them; but those which surround salient ones *curve round them*, preserving their continuity. At an angle of the latter kind too, crested or plume-shaped *interior* fringes are seen—"Grimaldi's crested fringes," as they are called, from the name of Father Grimaldi of Bologna, who first described (in 1665) these curious appearances. If a re-entering angle, however, be very acute, the external fringes which border its sides on approaching the angular point curve outwards, cross one another, and run out both ways into the shadow in elegant curves of a hyperbolic form. Nothing can be imagined more singular and *bizarre* than the appearance of shadows cast on a screen in this manner by a variety of minute objects of different shapes,—needles, feathers, lace-work, locks of hair, &c.; and as their observation, as we have shown, is exceedingly simple and easy, we earnestly recommend them to the attention of our readers—a bit of looking-glass, or, still better, a polished metallic reflector, a hole in a window shutter, a lens of an inch focus, a screen of white paper, and a sunny day, being all the requisites.

(108.) When the image of a small circular aperture (as a pin hole) is thrown on the screen, it is seen as a small round disc, highly coloured, the colours varying as the screen is approached from a distance to the hole—presenting in regular succession the tints of the reflected colours of thin plates described in a former part of this article, beginning with the first white: or, if the illumination be effected by homogenous light, alternate gradations of light from brightness down to total obscurity, and thence through an alternate succession of light and darkness. Around the central spot, too, coloured rings are formed, the tints of which vary in dependence on those of the centre. When the light is transmitted through two holes side by side, and very near together, besides the rings belonging to each, a set of *intersectional* coloured streaks is formed, straight if the holes be equal, hyperbolically curved if unequal. With three holes forming an equilateral triangle, or with a still greater number *arranged with perfect regularity* (as in machine-stamped paper in patterns), an endless variety of elegant and pleasing appearances will be witnessed. To see them to the greatest advantage, a magnifying glass should be used, *placing the eye in the place of the screen, and looking through the glass at the fringes and images of the holes as if they were real objects in its focus.*

(109.) When the system of apertures examined consists of a great multitude of exceedingly narrow parallel slits, precisely equal and equidistant, they constitute what is called a "diffractive grating," and present very curious, and in some cases brilliant, phænomena, which are best viewed by placing the eye close behind the grating. The luminous point (which appears colourless) is then seen accompanied laterally and on either side by a succession of highly coloured spectra, arranged in a line passing through it, and with their lengths directed along that line; their colours, unlike those of the fringes (which are composite) are the pure unmingled hues of the prismatic spectrum: even more vivid (if the grating be delicately executed) than the best spectrum which can be formed by refracting a sunbeam through a prism; and exceedingly remarkable in another respect, viz., that the proportional lengths of the coloured spaces in each, instead of depending, as in the case of the spectrum formed by a prism, on the nature of the particular medium of which the prism consists, is independent of any such consideration, and determined solely by the proportion between the wave-lengths corresponding to the colours of the rays. They are, therefore, what may be called *normal spectra.* So pure and undiluted indeed are their tints, that by the aid of a magnifying telescope the "fixed lines," so often above referred to, may be seen in them, and thus the wave-lengths, corresponding to the most conspicuous of these lines, ascertained with great precision. The violet ends of all the spectra are nearest to the central point, and the more distant spectra longer than the nearer, so that at length they overlap and confuse one another by the intermingling of the red end of one with the violet of that next in order.

(110.) If the apertures of which the grating consists be formed by removing with a graver portions of an opake varnish covering a glass surface, spectra exactly similar are seen accompanying the image of the luminous point reflected on the anterior surface of the glass from the polished portions laid bare. The same is observed, and with far more brilliancy, when a highly polished surface of metal is furrowed in equidistant parallel grooves by a graver or diamond point (which destroys the polish of those lines), and if the metal be hardened steel, the furrows so formed are transferable by violent pressure to the polished surface of a softer metal, which then in its turn exhibits similar appearances, and thus are produced the "buttons" above spoken of. Mother-of-pearl, too, which consists of exceedingly thin layers of calcareous matter superposed, and agglutinated or otherwise held together; when ground and polished, has these layers, which lie very little oblique to the general surface, torn up at their edges, where they crop out; which remain rough and unpolished, however brilliantly polished the general surface. The polished surface, therefore, is lined all over with almost exactly equidistant, exceedingly minute, non-reflective grooves, and when, if held close to the eye, a candle is seen in it reflected, its image accompanied with two lateral and very vivid spectra of similar origin, and an impression of the surface taken on black sealing-wax presents the same phænomenon.

(111.) When, as occasionally happens, the eyes are suffused with a nebulous film (due to the presence in the lacrymatory secretion of extremely minute globular particles of equal size), the image of a candle in a dark room some feet distant is seen surrounded with two or three broad circular halos of rainbow colours alternately ruddy and green. Similar halos are formed round the candle when viewed through two pieces of clear glass between which has been placed a little oil mixed with the delicate powder of the common puff-ball or *lycoperdon*, reduced to a thin even film by pressure and gently rubbing them together. In this case they are much more vivid and beautiful, the tints being those of the colours of thin plates beginning from the centre, only more dilute, so that it is difficult to discern more than a feeble indication of the fourth ring. Their diameters, however, unlike those of the coloured rings figured in Fig. 6, increase in arithmetical progression, or nearly so; that of the first or smallest, reckoned to the minimum of illumination, being $21^\circ\ 36'$ or thereabouts. They owe their origin to the exceeding minuteness, uniformity of size and sphericity (or at least circularity) of outline (if flat discs) of the spores of this fungus.

(112.) The explanation of these and of other phænomena referable to the head of diffraction, turns on two considerations, one of which may be regarded as a theoretical postulate (founded, however, on the analogy of sound), viz., that if a portion of a luminous wave be intercepted, the non-intercepted undulation spreads laterally into the dark space beyond, diminishing, however, in intensity as the lateral deviation of the *ray* (or perpendicular to the wave) from its original rectilinear course increases:—the other, a natural consequence of the mode in which a wave is propagated, viz., that every point in the surface of the non-intercepted portion may be regarded as the origin of a new wave spreading out spherically in all directions from that point as a centre; only with this proviso, that all such secondary waves start, each from its own origin, at the same precise instant of time and in *the same precise phase of its undulation.* And this, be-

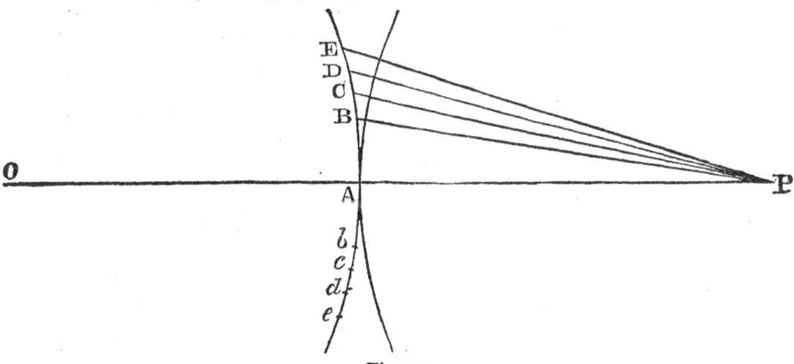

Fig. 10.

cause all belong to one wave surface, and are therefore necessarily coincident in time and identical in phase.

(113.) To show how this last consideration affects the question of diffraction, let us suppose a point P on a screen illuminated by light emanating from a single bright point O (Fig. 10), from which is propagated a series of equidistant spherical waves corresponding to light of any one refrangibility, and therefore distant from each other by one entire undulation of such light (say, to fix our ideas, a 50,000th of an inch). If O P be joined, intersecting the surface of any one such wave, at a given distance, O A from O in A; P A will be the shortest line that can be drawn from P to that surface. Suppose, now, we take on either side of A a series of points B, *b*; C, *c*; D, *d*, &c., progressively more distant (by pairs) from P than A is, by 1, 2, 3, &c., hundred-thousandths of an inch, or semi-undulations of the light under consideration; and let the whole figure be conceived as turned round on O P as an axis. Then these points will mark off on the spherical surface of the wave, a central circular area (call it the area A), and a series of concentric rings or rather zones of the waves (call them in succession B, C, D, &c.), surrounding it, like those represented in Fig. 7, from every point in each one of which the light sent to P will reach it in more or less discordance of phase, with that which reaches it from the next in succession. Thus if all the vibrations propagated from the central circle (A) arrive at P in a phase of compression, all these *simultaneously* reaching it from the zone (B) will arrive in a phase of expansion, all from (C) again in one of compression, and so on alternately. Now if the distance A P of P from the wave be anything considerable, suppose a few feet or even inches, it will be *enormously* great in proportion to one semi-undulation or 100,000th of an inch, the length by which the distances P A, P B, &c., differ from each other, and in that case it is very easy to show by geometry, that the successive areas (A), (B), (C), &c., are almost exactly equal. Were these areas *rigorously* equal, and were moreover the vibrations (propagated as they are from them to P *more and more obliquely* with respect to the general surface of the wave at their points of emanation) all of equal intensity, it would follow therefore that the otality of the movement propagated to P from (A) would be precisely opposed and destroyed by that from (B), that from (C) by that from (D), and so on; so that an ethereal molecule at P would in effect be agitated by no preponderating movement, one way or another, and there would be no illumination on the screen at P. Inasmuch, however, as the vibrations diminish in intensity as they are propagated more obliquely, and as the areas (A), (B), (C), (D), &c., are, though very nearly, yet not rigorously equal, this mutual destruction in the case of each consecutive pair is only partial, and the point P will be agitated by the sum of all these outstanding excesses (taken in pairs) from the centre outwards; which, though excessively small individually, in virtue of their immense number make up a finite sum. And as the same is true for each point of the screen (if spherical, and therefore everywhere equidistant from O), the whole of its surface will be equally illuminated: if plane, very nearly so, in all the region around P.

(114.) A very singular consequence follows from this reasoning, and one admirably calculated to test its validity, and the soundness of the theory it relies on, by experiment. There can be no better presumptive evidence of the truth of a physical theory than its enabling us to predict, antecedent to trial, a result in direct contradiction to what mankind in general would consider as the obvious conclusion of common sense founded on all *ordinary* experience. This is the case in the present instance. Since the total illumination of one point P on the screen is only that due to the undulations which remain outstanding after the mutual destruction of by far the greater proportion of those propagated from the zones (A), (C), (E), &c., (the *odd* zones, reckoning (A) as No. 1), by those emanating from the even ones (B), (D), (F), &c., it follows that if all the even zones could be entirely suppressed or rendered ineffective, the illumination at P would be prodigiously increased, and that even the obliteration of a few of them would produce a very material augmentation of brightness at that point. In other words, that by *stopping out* a large proportion of the luminous rays passing through a circular aperture from a bright illuminating point, the illumination of the central point of the image of such aperture thrown on a screen at a certain distance behind it, may be made to *exceed by many times what it would be were the whole aperture left open.* This strangely paradoxical result is stated by M. Billet* to have been experiment-

* Billet, *Traité d'optique Physique*, 1858, ii. § 55, by far the fullest *résumé* of that subject hitherto published; only too little explanatory, and sadly deficient in facility of reference. It deserves a good index.

ally verified by M. Fresnel (to whom its suggestion is due), and more recently by M. Billet himself, who by merely interposing (concentrically) between the luminous point and the centre of the screen, a small *opake* annulus exactly corresponding to the calculated dimensions (for red rays and using red light) of the first even ring (B) obtained an illumination at P estimated at five times that when no obstacle was interposed.

(115.) By way of showing the kind of explanation these principles afford of some of the simplest and easiest cases of diffraction (for their calculation is for the most part very complicated in its details, though simple enough in its principles); let us suppose first the case of a screen illuminated by a minute radiant point O through a small circular aperture, and consider only the illumination of the central point of projection on the screen, or of P in our figure. Suppose P to approach the screen from a very great distance—so great that the *difference of its distance from the centre and either edge of the aperture* shall be less than a semi-undulation of the light considered (say 100,000th of an inch). Then the undulations from every part of the aperture will reach P in phases more or less accordant with each other, and P will therefore be more or less illuminated: and, P still approaching, its illumination will increase till it attains such a distance that the difference in question exactly equals a semi-undulation. In this case the portion of the wave transmitted corresponds precisely to the whole of the central circle (A) of our system of wave-zones above discussed, and we have here the greatest possible amount of concordant and no discordant rays, and the illumination will be a maximum. But if P approach nearer, the difference in question will be greater than a semi-undulation, and the portion of an aperture near the edges will send rays to P more or less in *dis*cordance with those from the centre, and these will destroy a portion of P's illumination. P approaching this will go on till the difference amounts to an entire undulation (1-50,000th); that is to say, till the aperture extends (as respects the new situation of P) over the whole of the *two* first zones (A) and (B), of which the second (B) destroys almost the whole of the light from (A) (being equal in area, and differing very little in obliquity). Here then the illumination at P will be *nil* or very small. P still approaching, the third zone (C) (which sends vibrations in unison with (A)) will begin to be included within the limits of the aperture, and the illumination will again increase to another maximum, viz., when the three, (A), (B), (C), are just included; thence again it will diminish to a degree of obscuration not *quite* so complete as before; and so on. Thus as the screen approaches the aperture, its central point, after attaining a maximum of illumination, will suffer a succession of eclipses or obscurations gradually less and less complete, with intermediate recoveries, just as we have seen above, is really the case. When the light is not homogeneous, the different coloured rays having different wave-lengths, the obscurations of one colour will not correspond to those of the others, and thus will arise a succession of colours at the central point of the screen, agreeing with those there described.

(116.) Take now the case of the exterior fringes, when the shadow of a broad straight-edged body, as a ruler, is thrown on a fixed screen at a considerable distance behind it. Suppose P first placed exactly at the edge of the geometrical shadow. In that case, the view of exactly half of each of the concentric wave-zones (A), (B), (C), &c., will be intercepted, and P will therefore receive from the remaining halves just half the amount of *luminiferous* agitation it received when opposed to the whole wave, viz., half the amount of concordant and half of discordant undulation. Its intensity of *illumination* will therefore be *one-fourth* of that when the ruler is altogether removed.* Now, suppose the ruler withdrawn gradually, and *laterally*, so as to disclose to the view of P successively, 1st, the whole of the central zone (A) of the wave surface; 2dly, the whole of the two first zones (A), (B); 3dly, the three first, (A), (B), (C), and so on. It is very evident then, on merely casting our eyes on Fig. 6, (p. 295), and imagining a line drawn through the common centre of all the circles to be removed parallel to itself, step by step, so as to become in succession a tangent to the 1st, 2d, 3d, &c., circles; that in the first step of its removal it will disclose to P all the remaining half of the central area (A), which sends to it undulations *concordant* with those by which P is already illuminated, but less

* The effect on the retina is estimated, not by the simple *momentum* or *velocity* of the impulse communicated by the vibration, but by the "*vis viva*," "energy," or "work done," which is proportionate to the *square of the velocity* of movement. In this the undulatory doctrine of light agrees with the theory of sound.

than half of the second (discordant), still less of the third, &c., so that on the whole there will be a preponderance of the concordant undulations so introduced, and P will be *more strongly* illuminated than before. When removed one step farther however, since the newly introduced half of (B) the second zone almost exactly counteracts that of (A), the effect of the change will have its character decided by the proportional magnitudes of the segments of (C), (D), &c., disclosed, among which the preponderance is evidently in favour of (C), that is, of *discordant* undulation, so that by *this* removal of the shading obstacle the illumination of P will be *diminished;* and so on alternately. Now at each stage of these removals of the shading body, the edge of the geometrical shadow retreats farther and farther from P, or (which is the same thing) P is successively farther and farther outside of the edge of the shadow, becoming alternately more and less illuminated than at the actual edge. Here then we see the origin of the bright and dark external fringes exhibited in homogeneous light; and therefore by the very same reasoning, of the coloured ones produced by the successive overlapping of those formed by several coloured rays to each of which corresponds a different breadth of fringe; that for the red being broadest and for the violet narrowest.

(117.) *The twinkling or scintillation of the stars* partakes so far of the nature of a phænomenon of diffraction, as that it depends for its origin on the mutual interference of discordant rays arriving at one instant, but *by different routes*, on the same point of the retina of the eye; and which, therefore, do not interfere with or enfeeble one another in any part of their previous course. The image of a star on the retina is formed by the union in a focal point of the whole bundle or *pencil* of parallel rays contained within a cylindrical space or column, having the circular opening of the pupil for a base or section, continued through the whole atmosphere, however far it may extend. Now the air, though a very feebly refracting medium, has still a certain amount of refractive power, and *that* a variable one, depending on its density, temperature, and moisture; and corresponding to the degree of this power is the velocity with which it is traversed by the luminous undulations. Now; however the density, temperature, and moisture of the lower and upper regions of the air may differ; if throughout the whole extent of this column it were perfectly uniform in these respects, *at every point of each cross section* of it (however it might differ in different sections) all the rays traversing its length from the star to the eye would have their undulations *equally* retarded by the aerial medium: and therefore all the rays belonging to any one wave setting out at the same instant of time from the star would reach the focal point on the retina at the same moment; such being the condition which determines the focal point of a lens. But if the air in one side of the column should for any considerable distance along it be slightly different in these respects from that in the other, the undulations transmitted along that side would be differently retarded from those along the other, and would not arrive on the retina at the same instant. The one portion, then, on its arrival would meet there, not the other portion of the same wave to which it originally belonged, but one in advance or in arrear of that by either a whole, a half, or any part of an undulation, or any number of such, according to the extent of the difference in the quality of the aerial contents of the column. Suppose, for instance, the light from the two halves of the column to differ in their time of arrival by 1, 3, 5, or any odd number of semi-undulations of the most luminous or the yellow rays; these then would interfere and totally extinguish each other, and the apparent light of the star would undergo a great obscuration, assuming at the same time a hue complementary to yellow; *i.e.*, dark purple: and so for other rays. Now the constitution of the air is so irregular—such a perpetual mixture of masses of it, differing in temperature and moisture, is continually going on under the influence of wind-currents, that such differences as above supposed must be almost constantly in progress, even within the narrow space of a column no wider than the pupil of the eye, much more in that corresponding to the aperture of a small telescope. The scintillations, with their accompanying changes of colour, are beautifully seen through an opera-glass (*not binocular*), especially if somewhat out of focus, in which case the colours and the darkness are seen, as it were, to run over the circular disc into which the image is dilated in a very singular and capricious manner. If a small circular motion be given to the glass, so as to make the image of the star (when in focus) describe a circle, this will be seen *as* a luminous circle (as when a burning stick is whirled round, forming a ring of light), and in the circle so formed spaces highly coloured—red, blue, or green, as well as dark and bright—will be seen, corresponding to the state of the image at the instant it occupied those spaces, forming an easy, pleasing, and very interesting experiment.

(118.) An effect, the precise parallel to the scintillation of the stars, might be produced, affecting the ear instead of the eye, by sounding together two strings, at first exactly in unison, and then very slightly increasing and diminishing alternately the tension of one of them, thus producing a succession of beats, not regular as in the case of two strings permanently differing by a minute interval of pitch, but capricious in their succession.

LECTURE VIII.

ON LIGHT.

PART III.—DOUBLE REFRACTION—POLARIZATION.

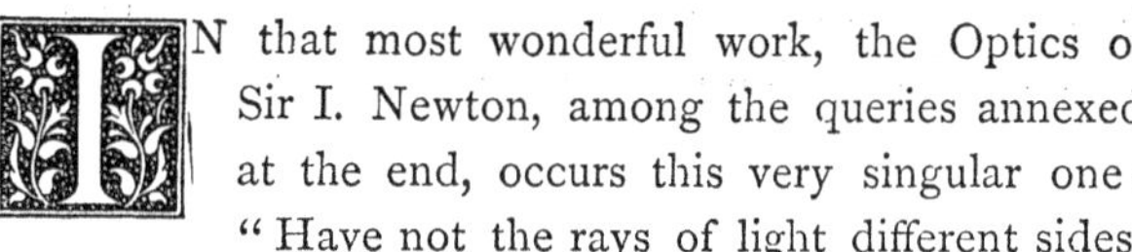

IN that most wonderful work, the Optics of Sir I. Newton, among the queries annexed at the end, occurs this very singular one: "Have not the rays of light different sides, endued with different original properties?"* The conception intended to be conveyed, as further illustrated by Newton himself, embodies that abstract notion of polarity which Dr Whewell in his "Philosophy of Inductive Science," expresses by "opposite qualities in opposite directions," or, as we should prefer to say, for this purpose, "different qualities in different directions," with reference (that is) to surrounding space and the objects therein situated. The same form of the general conception, as regards light, which Newton employed to designate the very same peculiarity in its habitudes, was adopted by Malus in his first announcement of the re-

* "Optics," Book iii., Query 26. 4th Edition.

markable discovery which introduced the term POLARIZATION into optical language. "We find," says he, "that light acquires properties which are relative only to the sides of the ray—which are the same for the north and south sides of the ray" (*i.e.*, of a *vertical* ray), "using the points of the compass for description's sake only, and which are different when we go from the north and south to the east and west sides of the ray." The polarization of light has in fact been an integral part of the science of optics (wanting only a name to designate it) ever since this suggestion of Newton, who derived it from the contemplation of one of what Bacon calls "instantiæ luciferæ," *luminiferous instances*, exhibiting the property or "nature searched after" "in an eminent manner," or in its clearest or most manifest form; and who described with the utmost clearness and precision the phænomenon in which its manifestation consisted in the special case before him.* We shall, therefore, approach the subject from Newton's point of view, choosing for our illustration the very phænomenon which led him to the singular conclusion embodied in his query.

* The same phænomenon is described, and with no less clearness and precision, by Huyghens, in his admirable work, "Traité de la lumière," published in 1690, fourteen years before the Optics of Newton—and from that epoch, or from 1678, when that treatise was communicated to the French Academy, must date the *discovery* of the polarization of light *as a fact.* Huyghens, moreover, *correctly* attributed it to a peculiarity impressed on the vibrations of the ethereal medium. But the picturesque phrase of Newton embodies the idea in a form easily apprehended, while it seems to have floated rather vaguely before the mind of his great predecessor, not so much as a general attribute, but as a specialty limited to the case in question.

(120.) As we have already stated when speaking of refraction generally, a ray of light incident on any transparent crystal (certain specified classes of crystals excepted) is subdivided by refraction into two distinct rays, pursuing different paths within the crystal, and of course emerging from it at different points, and so, of necessity, reappearing, not as a single, but as two distinct and separate rays, each pursuing its subsequent course independent of the other through space. Of these, when traced, one is found to have been refracted in the plane of its incidence on the transmitting surface, and according to the ordinary simple "law of the sines" already explained. It is therefore said to be *ordinarily* refracted, and it is called the *ordinary ray*. The other, except in special cases, deviates after refraction from the plane of its incidence, more or less according to the situation of that plane with *respect to the faces of the crystal:* and, moreover, in respect of the amount of its flexure, does not conform to the simple law of the sines, but to a rule much more complex in its expression, called the *law of extraordinary refraction;* this ray being designated as the *extraordinary ray*. Such is the case if the original, incident ray be one directly emitted from the sun, a candle, or any *self-luminous* body. But if in place of such a ray, we employ either of the two rays so separated as above described, for transmission through a second crystal of the same kind, the result will be very different. If it fall upon such second crystal in the same manner, according to the same angles and with the same relative situation as to its plane of incidence with the sides of the crystal, as in the case of the original incident ray, it will *not* be further subdivided, but refracted *singly: if an ordinary ray, ordinarily; and if extraordinary, extraordinarily.* Its refraction will also be single, if the second crystal be turned round *on the ray as an axis* exactly through a right angle; but in this case the second refraction, if an ordinary ray have been used, will be *extraordinay*, and *vice versâ.* In every intermediate situation of the second crystal, it will be subdivided into two, the one ordinarily, the other extraordinarily refracted, *but the two fractions will be found to differ in relative intensity:* generally speaking, the more according as the conversion of the second crystal has been through a less angle from its first position, and they are equal when the angle of conversion is 45°, 135°, 225°, or 315°, *i.e.*, exactly half-way between the rectangular positions of the crystal.

(121.) All these particulars are easily and elegantly exhibited by means of two crystals of the mineral called Iceland spar (crystallized carbonate of lime), a mineral of perfect and colourless transparency, which, if fractured, will always separate itself along its three "planes of cleavage" (which in this mineral are singularly distinct and palpable) into forms whose type is the obtuse rhomboid (whose six faces, all equal and similar rhombs of 101° 32′ and 78° 28′, are united three and three, by their obtuse angles, at the opposite extremities of a line called the axis of the rhomboid, the shortest that

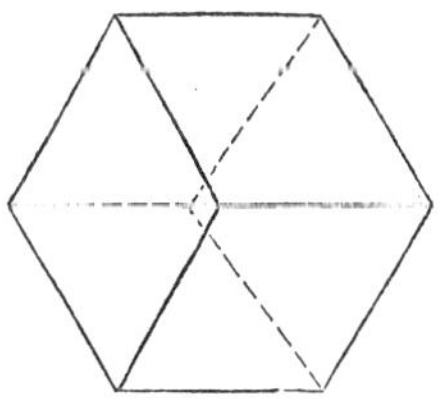

Fig. 11.

can be drawn across it, and to which it is symmetrical, as shown in the preceding figure).

(122.) If such a rhomboid be laid down on an ink-spot on white paper, or, still better, on a small pinhole in a plate of metal, and held up to the light, the ink-spot, or the dot of light, will appear through it doubled: the two images being separated, in the direction of the shorter diagonal of the face through which they are seen, by an interval of about one-ninth part of the thickness seen through. Thus, if over the first rhomboid another of equal thickness be laid *conformably* (*i.e.*, so that all the faces of the second shall be parallel to the corresponding ones of the first), the only effect will be that the apparent separation of the two images will be doubled, just as if a single crystal of double thickness had been used. But if from this position the upper crystal be turned slowly round in its own plane upon the lower, kept firm; two other images will make their appearance between the former, at first very faint and almost close together, but, as the rotation of the upper crystal continues, gaining strength (while the others grow fainter) and opening out from each other in a direction transverse to the line of junction of the first. When the angle of rotation attains 45°, four images are seen of equal intensity, after which the two first grow fainter, and at 90° vanish,—the whole of their light having passed into the other two—and so on alternately. When the upper rhomboid has made an exact semi-revolution on the other, the image is single, and contains the whole

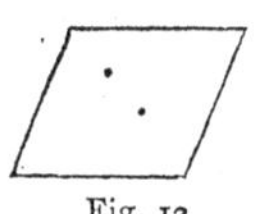
Fig. 12.

of the incident light. In this case, the ordinary ray has been refracted ordinarily, and preserves its situation; the extraordinary extraordinarily, but its displacement by the second refraction being exactly equal and opposite (in consequence of the now reversed position of the refracting rhomb) to that by the first, it is brought to coincidence with the other, and the two united form one image.

(123.) The opposite sides of a rhomboid being parallel, both the ordinary and extraordinary rays after transmission emerge parallel to the incident ray, by a necessary consequence of that general law of retro-version, in virtue of which a ray of light, whatever path it may have pursued from one point to another, can always retrace that path; the opposite faces being symmetrically situated with respect to the axis. And the same is true for a parallel plate of this or any other crystallized substance artificially cut and polished, whatever be the position which such plate may have held in the interior of the crystal from which it is cut. Now it is found, by cutting from rhombs of Iceland spar parallel plates in various directions, that there is one through which a ray of ordinary light can be transmitted perpendicularly without being divided into two. This is the case when the faces of the plate are at right angles to the line above designated as the *axis* of the rhomboid. And generally that a ray which within the crystal pursues a path parallel to this axis, will emerge from it single, whatever be the situation of the surface of emergence. The axis, then, is *a line of no double refraction*, and in the case of the substance in question (or of any crystallized body whose *primitive* form is the acute or obtuse rhomboid, the regular hexagonal prism, and some others, comprising all those primitive forms which can be described *as symmetrical to one line and to one only*) it is the only direction endued with this property. And on the other hand, the amount of the double refraction or the angular separation of the two rays into which the incident ray is divided, is greatest when they lie in a plane perpendicular to this axis. On account of these properties, the line in question is sometimes called the *optic axis* of the crystal.

(124.) If a crystal of Iceland spar, or any similar body, be cut into the form of a prism, in such a manner as to have its refracting edge parallel to its optic axis, neither of the two refracted rays will emerge parallel to the incident one, or to each other. They will diverge, including an angle between them, greater as the refracting angle of the prism is greater, exactly as if the medium had two different refractive indices. And in this particular case both refractions follow the ordinary "law of the sines," and there is no deviation of either ray from the plane of incidence. And what is extremely remarkable, not only the refractive indices, but the dispersive powers of the two refractions differ, in some cases widely, so as to give two spectra (when a sunbeam is refracted) of very different lengths. In the case of Iceland spar, the respective refractive indices for the ordinary and extraordinary ray are 1·654 and 1·483. It is in consequence of this great difference that the two images of a point, seen through a rhomboid of the mineral, appear unequally raised above their natural level; that seen by the ordinarily refracted rays, appearing nearer the eye than the other.

(125.) By the employment of such a prism as here described, it is easy to insulate either the ordinary or extraordinary refracted ray, and examine it separately. Suppose, for instance, the latter to be stopped by a screen, and the former only allowed to reach the eye. If before doing so it be made to pass through a second such prism, whose refracting edge is parallel to that of the first, it will be refracted *singly* and *ordinarily*: if the edge be held perpendicularly to that of the other, then, *singly* but *extraordinarily*. In every intermediate position the image will be doubled, more of the light passing into the extraordinary image, and less into the ordinary, according as the angle at which the edges cross is increased from 0° to 90°; and at 45° of inclination the light is equally divided between the two. This, it is obvious, could not be if the ray were indifferently disposed with respect to surrounding space. There subsists in it a difference of properties depending on situation—a difference *analogous to* that between a square rod and a round one. It has acquired *sides* in its passage through the crystal, which it preserves in its subsequent course through space till it meets some body whose action on it may bring their existence into ocular evidence. It would seem almost as if light consisted of particles having polarity, like magnets; and that in its passage through a doubly refracting substance these came to be arranged, half with their poles in one direction (as regards the sides of the crystal), and half in a direction at right angles to the former. And this is the way in which Newton conceived it, as he himself distinctly states (Opt., Query 21); and as it necessarily must be conceived if the corpuscular theory were to be adopted. How it is explained on the undulatory hypothesis, we shall presently see.

(126.) It was while gazing one evening in 1808, through such a prism of Iceland spar as we have just described, from his study in the Rue d'Enfer at the windows of the Luxembourg Palace in Paris, that M. Malus, at that time engaged in studying the law of extraordinary refraction in this body, happened to notice that the reflexion of the sun on a window of the palace disappeared from one of its images, in a certain position of the prism, and from the other when held at right angles to that position; while in the intermediate situations, the glare was visible in both images, unequally divided, however, between them, except when held in a situation exactly intermediate, or at 45° from its first position:—in a word, that the light reflected from the window had acquired precisely the peculiarity which would have been impressed on it by previous transmission through a similar prism. To this peculiarity he gave the name of POLARIZATION, and light so affected has ever since been said to be POLARIZED.

(127.) *Total and partial polarization of light by reflexion.*—The angle at which a ray of light must be incident on glass that the reflected ray may acquire this property is 56° 45′ from the perpendicular, or at an inclination of

33° 15′ to the surface, and in this case the polarization is *complete*, or the whole of the reflected light has acquired the property in question.* If at any other obliquity; it is only *partially* polarized, or a portion only of the reflected light has acquired it. How this portion is to be distinguished and separated from the *unpolarized* portion, we shall presently explain. Suffice it here to observe that this latter portion bears a greater proportion to the whole reflected beam, as the angle of incidence deviates more from that above specified (which is called the *polarizing angle*). The plane in which reflexion has been made is called the plane of polarization; and two rays which have undergone reflection at the polarizing angle in planes perpendicular to each other, are said to be *oppositely polarized.*

(128.) The angle of incidence 56° 45′ has this peculiarity—that if we consider the directions subsequently pursued by the two portions into which a ray so incident on glass is divided, the one pursuing its course by reflexion in the air, the other by refraction within the glass, these two directions include a right angle as in the figure overleaf, where A C is the incident, C B the reflected, and C D the refracted rays, at the surface of a glass P Q. When the angle A C P or B C Q is 33° 15′ Q C D is 56° 45′, and D C B is a right angle. The law of polarization *so announced*, as Sir David Brewster has shown, is general,

* In point of fact the differently coloured rays are not all polarized at exactly the same angle, so that this is rigorously exact only for homogeneous light. But the difference is so trifling that it is purposely here kept out of view.

and extends to all media, whatever their refractive powers. Thus, for water, the polarizing angle is 53° 11′, and for diamond 68° 6′—numbers concluded from this simple rule (equivalent to the geometrical property above

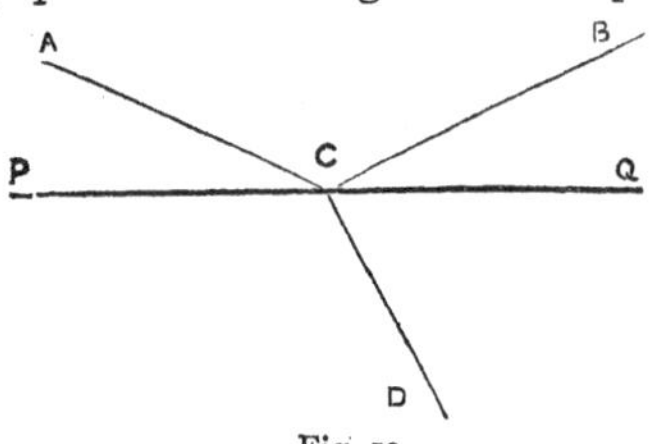

Fig. 13.

stated) that *the index of refraction is in all cases the tangent of the polarizing angle.*

(129.) If a ray of light so polarized by reflexion be received on such a prism as is above described, held with its refracting edge (*i.e.*, the optic axis of the crystal) at right angles to the plane of reflexion, or parallel to the reflecting surface, it will entirely pass into the ordinary, or most refracted image; *vice versâ*—if the prism be turned round 90°, so as to have its edge parallel to the plane of reflexion (*i.e.*, of polarization), wholly into the extraordinary. The same interchange will of course take place if the prism be held immovable, and the reflecting glass turned round, so as to change the plane of reflexion, and thus we perceive that rays *oppositely polarized* are distinguished by the characters of passing *entirely* into the one and entirely into the other of the two images, when so refracted. A beam of light partially polarized may be regarded as a mixture of two portions, the one wholly polarized, the other wholly unpolarized; and a beam of *unpolarized* light may, conversely, always be regarded as a mixture of two equal rays oppositely polarized, in *any* two planes at right angles to each other.

(130.) If a ray reflected from any medium at the polarizing angle (and therefore wholly polarized) be received on a second surface of the same medium at the same angle of incidence, *and in a plane coincident with* that of the first reflexion, it undergoes partial reflexion just as an unpolarized ray would do, and both the reflected and refracted portions retain their polarization. But if the plane of the second incidence be at right angles to that of the first, *no portion of the light is reflected*, but the whole passes into the refracted ray, retaining its polarization,—just in the same manner as, had it been incident on our doubly refracting prism held with its edge at right angles to its plane of polarization, it would have wholly passed into the extraordinary ray. *Vice versâ*, if the ray extraordinarily refracted by such a prism be received on a glass plate at the polarizing angle of incidence, no reflexion will take place if the edge of the prism be parallel to the plate. Hence we are entitled to conclude that it is the very same property which is impressed on light in both cases, and that a ray polarized by reflexion differs in no respect from one which has received this property by passing through a doubly refracting crystal.

(131.) A ray partially polarized by reflexion at a greater or less incidence than that at which it would have been completely so, may be wholly polarized, or nearly so, by repeated reflexions at the same angle.

This, which might be concluded *à priori* by considering that the unpolarized portion differs in no respect from ordinary light, and is therefore susceptible of so receiving partial polarization, while the polarized portion retains its polarization unchanged by reflexion, is verified by experiment.

(132.) If a ray partially polarized in any plane be received on the doubly refracting prism already mentioned, with its edge perpendicular to the plane of polarization, the polarized portion will pass wholly into the ordinary image, while the unpolarized will be equally divided between the two. Thus the two images will be unequally bright. By turning round such a prism, then, till a position is found at which the contrast between the two images is most striking, this plane will be discovered, and the difference of their illuminations is the measure of the quantity of polarized light in the beam.

(133.) *Polarization of light by refraction.*—When light is incident on glass or any uncrystallized transparent body *at the polarizing angle*, the reflected portion (a small per-centage, not more than one-twelfth of the whole light) is wholly polarized in the plane of incidence, as already stated. The refracted beam (by far the larger portion), when examined in the mode just described, is found to be *partially polarized in a plane at right angles* to that of incidence, and the amount of polarized light which it contains to be precisely equal to that in the reflected beam. Thus we see that when light falls upon such a surface, the greater portion passes unchanged, while the other is divided into two equal portions *oppo-*

sitely polarized, the one being reflected, the other transmitted, and intermingled with the unpolarized part.

(134.) If a parallel plate of glass be used for this experiment, the same process is repeated at the hinder surface. An equal per-centage of the unpolarized portion is similarly divided between the reflected and transmitted rays, oppositely polarized, and as the transmitted polarized portion is, *ipso facto,* guaranteed from subsequent reflection at the polarizing angle, the total amount of polarized light in each of the two beams is nearly doubled. If behind this a second parallel glass plate be applied, the same process is again repeated on the remaining unpolarized portion, and so, by multiplying the plates, the whole incident beam is ultimately divided equally into a reflected and a refracted beam completely polarized in opposite planes. Such at least would be the case were the plates perfectly transparent and infinite in number; but as these conditions cannot be fully realized, the transmitted beam is never completely polarized, though enough so to afford a very convenient mode of viewing many optical phænomena. On the other hand, if the plates be truly plane and their surfaces exactly parallel, the reflected beam is wholly polarized, and as its intensity is nearly half that of the incident light, this affords an excellent mode of procuring a polarized beam available for purposes of experiment. A frame containing six or eight squares of good window glass laid one on the other, and backed by a sheet of black velvet, is one of the most convenient and useful of optical instruments, and will be frequently referred to in what follows as a "polarizing frame," a similar series set transparently being termed a "polarizing pile."

(135.) *Other modes of polarization.*—There are certain doubly refractive crystals, more or less coloured, which possess the singular property of absorbing or stifling in their passage through them, *unequally*, the two oppositely polarized rays into which they divide the incident light. Two bodies especially have been noticed in an eminent degree endowed with this property—the one a mineral occurring more or less frequently among the rocks of igneous origin, called the tourmaline, the other an artificial chemical compound, the iodo-sulphate of quinine. The former crystallizes for the most part in long prisms of many sides, terminated by faces, three of which belong to the primitive form, an obtuse rhomboid, whose axis is that of the prism itself. In consequence, like all crystals of this class, it is doubly refractive, and if artificially cut and polished into a doubly refracting prism, having its refracting edge parallel to that of the rhomboid, an object viewed transversely through it will appear double; provided the eye be held quite close to the refracting edge; but if ever so little removed, so as to look through a greater thickness of the substance, one of the images will be observed to diminish rapidly in intensity; and at a certain, usually very moderate, thickness to disappear altogether, as if extinguished by a deeper colour or a higher degree of opacity in the medium, the other remaining undiminished. The unextinguished ray is, of course, completely polarized, and *that* in the plane of the section of the prism at right angles to the edge. If, instead of cutting the crystal into such a prism, it be formed into a flat plate, with its faces parallel to the axis of the rhomboid; such plate will in like manner extinguish one of the pencils into which a ray incident perpendicularly on it is divided, allowing the other to pass; and the pencil so transmitted is completely polarized in a plane perpendicular to the axis of the plate. This property of a tourmaline plate renders it invaluable as an optical instrument, affording the readiest and most convenient means of procuring a polarized beam of light for the examination of crystals and other purposes. Its only drawback is, that this mineral is most commonly coloured with a strong tint of blue or green, which affects the colour of the transmitted light. Some specimens, however, while equally effective in destroying one of the refracted pencils, are yet but slightly tinged with colour as respects the other, which is therefore transmitted fully polarized, but with only a slight tinge of brownish yellow. The other substance, of late much resorted to for the same purpose, the iodo-sulphate of quinine, crystallizes in very thin scales like mica, of a purplish-brown hue, which in like manner polarize completely one half the incident light, which passes freely through them; the other half being extinguished. This curious property was discovered by Mr Herapath, who first formed the compound in question.

(136.) When two parallel plates of tourmaline cut from the same crystal in the mode above described, or two laminæ of Mr Herapath's quinine salt, are laid one on the other conformably (or with their axes parallel), the light polarized by one passes freely through the other; but if the one be turned round on its own plane, on the other, the intensity of the transmittted beam gradually diminishes, until the axes cross at right angles, in which position the combination is quite opake. A similar gradual diminution of light, up to complete extinction, takes place when a ray polarized by reflexion from glass, or in any other manner, is received on such a plate, made to rotate in its own plane. The effects are just what might be expected to happen if a flight of *flattened* arrows were discharged at a grating of parallel wires. Those only whose planes were parallel to the wires would be able to pass, and having passed one such grating, would penetrate any number of others placed conformably behind it, but not if placed transversely. This is a simile, not an explanation, but it conveys, though coarsely, to the mind a conception of the distinction between polarized and unpolarized light, not to be despised as an aid to the imagination.

(137.) If a "polarizing frame" of glass plates, such as above described, be laid down before an open window, and, the eye being held near it so as to embrace a large illuminated area or "field of view," a tourmaline plate be looked through, and turned slowly round, in its own plane before the eye; a position will be found in which the appearance of a dark cloud comes over the frame, extending over a very considerable visual angle; the central portion being completely obscure; and the darkness shading off at the borders by insensible though rapid degrees. Within this "polarized field," a vast variety of brilliant and beautiful optical phænomena, hereafter to be described, are very conveniently and elegantly exhibited. One effect is very striking. If instead of a black velvet backing, the glasses be laid on any bright surface, the printed page of a book, for instance—this, which, without the interposition of the tourmaline, cannot be discerned for the glare of light reflected from the glasses, becomes distinctly visible, and may be read with facility when that glare is taken off in the manner described. So, too, by looking through a tourmaline plate held transversely, on the surface of a pond, at the polarizing angle; the reflected light from the surface being destroyed, the objects at the bottom, the fishes, &c., are distinctly seen, though completely invisible to a bystander. So, too, by polarizing alternately in a vertical and a horizontal plane, the light of one or more lamps, night signals may be made, and a message transmitted, visible *and interpretable as signals*, to a distant spectator provided with a tourmaline plate, while a bystander not so provided, though he see the lamps, will have no suspicion that any such communication is in progress.*

(138.) *Polarization of the sky light.*—The light of a clear and *perfectly* cloudless blue sky is partially polarized in a plane passing through the sun, the eye, and the point of the sky examined. At each point in that great circle of

* I mention this to prevent a patent being hereafter taken out "for secret communication at a distance by means of polarized light."

the celestial concave, which is 90° remote from the sun's place, the amount of polarized light which it sends to the eye bears a very considerable proportion to the total illumination, amounting to nearly a fourth of the whole. At every other inclination of the visual ray to the direct sunbeam, the proportion is less, and in the neighbourhood of the sun, or of the point on the horizon directly opposed to it very small. When examined in a mode hereafter to be described, by the intervention of a tourmaline plate, and a crystallized lamina, this gives rise to a series of exceedingly beautiful and brilliant phænomena, productive of the greatest astonishment to those who learn for the first time by their exhibition, that totally different, and even opposite qualities, characterize different portions of an illuminated surface apparently so perfectly uniform and homogeneous. This effect is supposed to originate in the reflexion of the solar rays on the particles of the air itself, an explanation encumbered with many difficulties, but the best (indeed the only one) that has yet been offered.

(139.) *Polarization interpreted on the undulatory theory.*—According to any conception we can form of an elastic medium, its particles must be conceived free to move (within certain limits greater or less according to the coercive forces which may restrain them) in every direction from their positions of rest, or equilibrium. It by no means follows, however, that the nerves of the retina are equally susceptible of excitement by vibrations of the luminiferous ether (in which they may be conceived immersed) in all directions. In the case of sound, the tympanum of the ear which receives the impulse of the aerial medium, would appear to vibrate like the parchment of a drum, by the direct impact of its waves perpendicular to its surface. It is, therefore, sensible to such of the movements of the vibrating medium *only* as are in the direction of the sound-ray, and not at all to transverse vibrations. But if we conceive the nervous filaments of the retina as minute elastic fibres, standing forth at right angles to its plane, like the bristles of a brush (the reader will pardon the apparent coarseness of the illustration, which is only intended *as* an illustration of what may be, and no doubt is, a process of transcendent delicacy), immersed in the ether; it is evident that movements of the latter parallel to their direction *would not*, but that those transverse to it *would* tend to throw them into vibration, just as ears of corn would be little or not at all agitated by a straight and slender rod moved up and down between the stalks, or to and fro in the direction of its own length, but violently by a transverse horizontal motion of the rod.

(140.) Whatever be, at any instant, the motion of an ethereal molecule, it may always be resolved into two, one in the direction of the ray in the act of propagation, and the other in a direction transverse to it, in the plane of the wave surface. If the sensation of light be supposed to be produced by the former resolved portion, no account can be given of the phænomenon of polarization; such movement being equally related to surrounding space in all directions outward from the ray as an axis. The contrary is obviously the case with the other resolved portion. Suppose one end of a long horizontal cord fastened to a wall, the other held in the hand, and tightly strained. If a small vibratory motion cross-wise to the cord be given to the hand *in a horizontal plane*, an undulation *confined to that plane*, will run along the cord; if in a vertical one, then will the undulation be wholly performed in a vertical plane. If the propagation of a wave along a stretched cord be assimilated to that of a ray of light, the former of these cases will convey the idea of a ray polarized in a horizontal, the latter in a vertical plane. If the movement of the hand (always transverse to the cord) be not confined to any particular plane, but take place sometimes in one, sometimes in another, at all sorts of inclinations to the horizon—the undulation which runs along the cord in this case will convey the idea of an unpolarized ray. (According to Sir David Brewster, however, a partially polarized ray would, in this manner of viewing it, be assimilated to the case when the vibratory movement should neither be strictly confined to one plane, nor altogether irregular, but confined in its deviations from it to some angular limit less than a right angle.)

(141.) There is nothing to lead us to believe that the vibratory motions of the particles of material bodies, especially those in the state of gases in the act of combustion, in virtue of which they are luminous, are necessarily confined to any particular plane. Many thousands, or even millions, of vibrations in one plane may be succeeded by as many in any other, according to the direction and frequency of the shocks which give rise to them within an interval of time inappretiably short, and without prejudice to the continuous perception of the vibratory movement communicated to the ether *as light.* The act of polarization consists then in the *subsequent* arrangement, at some definite point in the line of progress of the ray, of all these vibratory movements, into parallelism with each other, or into a single plane from which they have afterwards no tendency (*per se*) to deviate. As the particles of crystallized bodies must be conceived to be arranged in definite lines and planes, it is easily conceivable that, whether among them, or in conjunction with them, the ethereal molecules may be confined in their vibrations to two particular planes determined by the internal constitution of the crystal and the incidence of the light; that a vibratory movement propagated into a body so constituted should *ipso facto* resolve itself into two such movements in these two planes (according to the general mechanical principle of the composition and resolution of motion); that it should be so propagated during its progress through the crystal; and that at its emergence into free space, each vibration should thenceforward subsist separately, there being nothing to change it. Again, it is no less conceivable that in these vibrations the molecules of the ether moving in one plane may be differently impeded by, or stand in a different connexion with those of the medium, from those moving in the other, and that, in consequence, their propagation of the movement may be effected with a different velocity, and thus give rise to a difference of refractive power, which, as we have seen, depends on the proportion of velocity of the light in and out of a medium. In the case of one set of vibrations, again, the propagation of the medium may be equally impeded or influenced in all directions of the ray—in which case a wave starting from any point in the surface would run out spherically within the crystal, while in that of the other the amount of obstruction might vary with the direction of the ray, and thus give rise to a wave running out *with different velocities in different directions* from its centre of propagation, and therefore *not spherically*.

(142.) To this conclusion, but without passing through the intermediate considerations which have led us up to it, Huyghens (who certainly had formed no conception of transverse vibrations) appears to have *jumped*, by one of the happiest divinations on record in the history of science, viz., that in the double refraction of Iceland spar, while the ordinary ray is propagated in a *spherical*, that of the extraordinary spreads from its point of origin at the surface of the crystal in an *elliptical* wave, the form being that of an *oblate spheroid* of revolution, having its polar axis parallel to the axis of the rhomboid, and bearing to its equatorial diameter a definite numerical proportion, viz., that of eight to nine (very nearly). Making this assumption, and laying it down as a principle (capable of demonstration), that the direction of a ray of light in such a mode of propagation is not that of a perpendicular to the surface of the wave at any point, but that of a line drawn from the centre of the wave to its point of contact with a plane, touching at the same time all the wave surfaces in progress, at the same time, through the crystal, which have originated in one and the same plane wave sweeping over its external surface (just as in the explanation of ordinary refraction given in our first part, in the case of spherical waves, in which case the latter line *is* perpendicular to the wave surface); he was enabled to explain every particular of the double refraction in Iceland spar, so far as the *direction* of the extraordinary ray is concerned, including its deviation from the plane of the angle of incidence, and its non-conformity with the ordinary law of the sines except in special cases. The results of his reasoning have been compared with experiment, with extreme care, by M. Malus, as already mentioned, and found exactly in accordance with fact. We cannot, of course, in an essay like the present, give any account of the special conclusions, or of the mathematical reasoning on which they are founded; which involve more geometry than the generality of our readers are likely to possess. But we can put into a very few words, and we think make readily intelligible, the main feature of the reasoning, that which determines the deviation of the extraordinary ray from perpendicularity to the wave surface, and from the plane of incidence.

(143.) Let A B C D represent a plane wave descending perpendicularly upon the upper surface E H (supposed horizontal) of a crystal of Iceland spar, of which E H M I is the principal section, or that cutting through both the obtuse angles of the rhomboid, and in which its axis lies.

The light then which this wave conveys will be incident perpendicularly on the surface E H, or in the direction of the lines B F, C G, and these lines continued to K and L, on the lower surface, would be the course of the *rays* B F, C G, supposing them to undergo the *ordinary* refraction. Considering now the *extraordinary;* suppose the portions E F, G H of the surface screened, and only the portion F G of the wave allowed to enter. This on striking the surface, will excite at every point over its whole extent a luminiferous vibration, which will be propagated

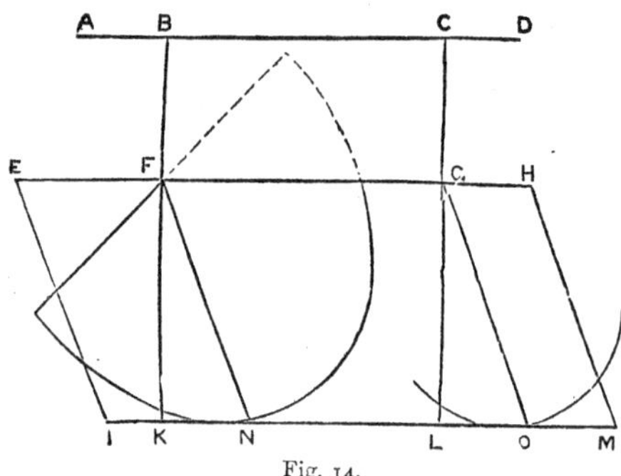

Fig. 14.

within the crystal in a spheroidal wave, having its shorter axis parallel to that of the crystal: and all these spheroids being equal and similar, the plane which touches them all, and which *is*, in effect, the extraordinarily refracted *plane wave within the crystal*, will advance parallel to the surface E H. Suppose it arrived at the other surface I M, and let N O be the points of contact of that surface with the spheroidal elementary waves whose centres are F and G at that moment. Then will N O be that portion of the posterior surface which will *first be struck* by the refracted wave. The portions beyond on either side, I N and O M, will *subsequently* receive the divergent undulations, which (as we have already explained) give rise to diffracted fringes bordering the shadows of the screens E F, G H. Thus we see that the space between N and O, and not that between K and L, will receive the full illumination from the aperture F G, which has therefore been propagated obliquely in the direction of the lines F N, G O, and not of the perpendiculars F K, G L.*

(144.) The deviation of the refracted ray from *the plane* of incidence, and from that of ordinary refraction, will be readily understood when it is borne in mind that whether at a perpendicular or an oblique incidence, a *plane exterior wave* is transformed by the extraordinary, as well as the ordinary refraction, into a plane *interior* one, and that the plane of incidence *of a ray* is perpendicular to both these planes. It cannot therefore contain the extraordinary refracted ray (which is a radius of the spheroid) without containing at the same time a normal to the elliptic surface of propagation at its point of contact with the interior plane wave, that is to say, unless it contain also the axis of the spheroid. In other words, the extraordinary ray will always deviate from the plane of incidence, unless in the case when that plane coincides with some one of the meridians of the spheroid in question.

(145.) Since there is no double refraction in the direction of the axis of the rhomboid, it follows that in that

* This is Huyghens's explanation, and the correct one.

direction the velocities of propagation of the ordinary and extraordinary rays within the crystal are the same, and that therefore, supposing the two corresponding undulations propagated from the same point in its surface to run out internally, the one in the form of a spherical, the other of a spheroidal shell, these shells will have a common axis, viz. :—the shorter axis of the spheroid, which will therefore wholly include the sphere, being in contact with it at the poles of the spheroid. *Cæteris paribus*, too, it is equally obvious that, when we come to consider different sorts of crystals possessing the property of double refraction, the intensity of this quality, or the amount of angular separation of the two refracted rays at the same incidence, will be determined by the greater or less amount of ellipticity of the spheroid in question. Should this ellipticity be *nil*, the spheroid will coincide over its whole extent with the sphere, and there will be no double refraction. This is the case with all crystallized bodies, whether mineral or artificial salts, whose primitive form is the cube. In some cases (comparatively rare ones), of which quartz or common rock crystal is an example, the spheroid is of the kind called *prolate*, or one formed by the revolution of an ellipse round its *longest* diameter, and is therefore wholly contained within the sphere. In these, then, the velocity of the extraordinary ray within the crystal is less than that of the ordinary; and the latter ray, which is the more refracted of the two in Iceland spar, is in such crystals the less so. On comparing different crystals, however, it is not found (which perhaps might have been expected) that the ellipticity of the spheroid in question is determined solely, or even principally, by the degree of obtuseness of the rhomboidal form of the crystal. It appears to be regulated far more by the chemical and other physical qualities of the material.

(146.) *Of the interference of polarized rays.*—The assimilation of a ray of light to a series of equidistant waves running along a stretched string, will afford a very clear conception of the interference of polarized rays. Suppose a vibratory movement *in a horizontal plane* to be communicated to one end of such a string, and to propagate along it such a series of waves, which will therefore all be confined to the same horizontal plane. If then a simultaneous movement, exactly equal and similar, *and in the same plane*, were communicated to a point in the string exactly half a wave breadth in advance of the point where the first series originated; each point in its length anywhere in advance of both these origins of movement would be always solicited by two equal and opposite impulses, the one of which would contradict the other, and in consequence it would remain at rest, and the two series of waves would destroy one another. If the origin of the two vibratory movements were distant from each other by a whole wave breadth, they would conspire to produce a double extent of vibratory excursion all along the string. All this is merely recapitulatory of what was stated, in Lecture VII., when explaining the general nature of the interference of rays. But it is evident that these conclusions only follow if the interfering vibratory movements are performed *in the same plane.* Supposing them performed in planes at right angles to each other, no such mutual destruction or reinforcement of movement can take place. Two movements at right angles to each other, communicated at the same instant to the same material molecule, combine, in virtue of the mechanical principle of the composition of motions, to produce a movement intermediate in direction; and can in no case destroy each other.

(147.) It follows from this, that if such be really the nature of the luminous vibrations and such the true explanation of the phænomenon of polarization—interference can only take place between rays polarized in the same plane—such complete interference at least as shall result in the extinction of both, in the manner above described. This conclusion is, happily, capable of being brought to the test of experiment, and the result is found to be in exact accordance with the *à priori* reasoning. The experiment is simple and direct. Let two small holes, or, better, parallel slits very near each other in a thin opake screen, be placed between the eye and a very minute and brilliant point of light; and viewed through a lens, as described in a former paragraph; so as to see the diffractive fringes. Now over the holes or slits let two plates of tourmaline of precisely equal thickness and in every respect similar be applied (to secure which conditions, the two halves of a single plate worked to exact parallelism and cut across, may be used). Then if the axes of these plates be parallel, in which case the light passing through both the apertures will be similarly polarized, the fringes will continue to be seen. If one of them be slowly turned round in its own plane till its axis comes to be situate at right angles to that of the other, they will gradually decrease in intensity and at length disappear altogether when this rectangularity is precisely attained. In the first case, then, the rays have interfered—in the last, not: while in the intermediate states a partial interference takes place, the more complete the nearer the axes are to parallelism. How *this* is operated we shall now proceed to explain.

(148.) *Circular and elliptic polarization.*—If we regard the vibratory movement of any single particle of an elastic medium in its most general mode of conception, we shall find that it may always be considered as capable of resolution into three rectilinear vibrations in three planes at right angles to each other, each going on *as if* the others had no existence: and its place in space at any instant will be had by estimating its distance on one side or the other of its neutral or central position (those of perfect equilibrium and rest), reckoned along each of the three lines in which these planes intersect (which, after the manner of geometers, may be considered as three rectangular axes, or co-ordinate lines), which it would have attained at that instant in virtue of each separately, and independent of the others. This is nothing more than the enunciation of one of the simplest of mechanical laws, that of the composition and resolution of motions. But the theory of movements propagated through elastic media (a theory far too elevated and intricate to admit of any explanation in these pages, and whose results the reader must take for granted) further teaches us that a vibratory movement once set up and steadily maintained, according to a regular law of periodicity in any one molecule of such a fluid will, sooner or later (according to its distance and situation), reach every other, which from that moment will be agitated by a vibratory movement precisely similar in its phases (though of inferior extent in its excursions) to the original movement, and *performed in the same period of time.* It matters not whether the medium be or be not equally elastic in all directions. This will affect the *rate of progress* of a wave through it in different directions, and by consequence the form of the wave, and the length of time that has to elapse before the molecule in question begins its vibratory movement; but once set up in any molecule, that movement will be maintained, so long as it is, so to speak, fed from behind—so long as successive waves continue to pass through it. In the theory of light, the eye being insensible to vibratory movements in the direction of the ray, we have only to consider those components of the motion which lie in a plane at right angles to that direction, and which, for the present, we will suppose to be that of the paper before us.

(149.) Let us consider, then, the kind of motion which an ethereal molecule will assume, under the influence of two such vibratory movements simultaneously affecting it, in directions transverse or otherwise inclined to each other, but both directions lying in one common plane, that of the paper, or of the wave-surface; each of which will therefore represent the vibratory movement proper to a ray polarized in a plane at right angles to our paper, and intersecting it in its line of direction. And first, for simplicity, let us suppose the two vibrations of equal intensity (*i.e.*, in both which the molecular excursions on either side of the point of rest are equal), and that their directions form a right angle with each other. Let A B and *a b*, fig. 15, represent two such lines of vibratory movement, C, *c*, their central points, or the positions of rest of the molecules when undisturbed, and C A, C B; *c a*, *c b*; their extreme excursions to and fro. The times of vibration being equal (which is an indispensable condition for the union into one of two distinct luminous rays: as a red ray, for instance, cannot interfere with a violet one), let each be supposed divided into the same number of equal parts (say 360). Then supposing the molecules to set out at the same instant from C and *c*, they will arrive at A *a*, respectively in 90 such units of time, will have returned again to C *c*, in 180; have reached B, *b*, in 270, and again returned to C and *c*, in 360. In so doing, however, their motions are not uniform, but most rapid when traversing the central points, and gradually retarded as they recede from these: so that in equal intervals of time the spaces traversed along the lines C A, *c a*, will be unequal. Then let the whole time (90) of describing C A, be divided into five equal times of 18 each, and suppose that at the end of the 1st, 2d, 3d, 4th, and 5th of these, the molecule has arrived at the points 1, 2, 3, 4, 5. It is demonstrable, then, that the several distances C 1, C 2, C 3, C 4, C 5, of these points from C will be to each other in the proportion of the Sines of 18°, 36°, 54°, 72°, and of 90°

or radius; and the same is true of *c* 1, *c* 2, *c* 3, *c* 4, *c* 5.

(150.) This premised, we are now in a condition to trace the movement of a molecule affected at once by both these causes of displacement. Let it be O. Then at the expiration of the first interval of 18 units of time it will in virture of the vibration parallel to C A, be carried to a distance equal to C 1 in a direction O P parallel to that line, and will be found so far from the

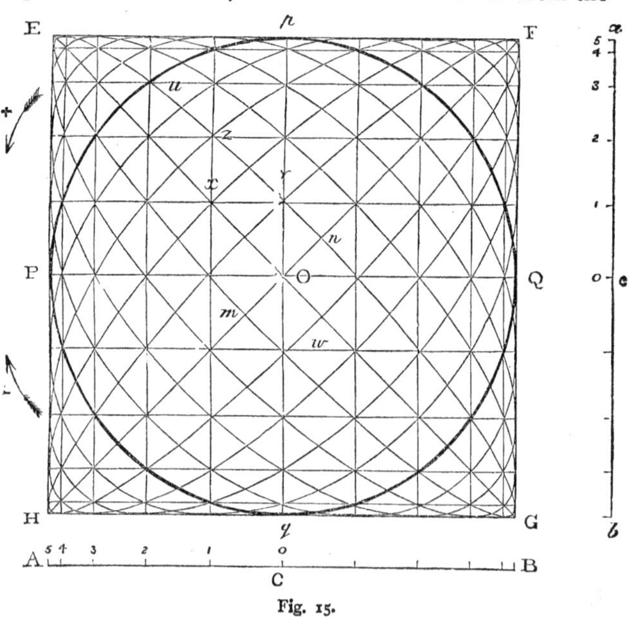

Fig. 15.

line O *p* parallel to *c a*. And in virtue of the other vibration similarly it will be found at the same time at the distance c I (=C I) from O P, or in the direction O p parallel to c a. In virtue of both movements, then, it will be found at X, the extremity of the diagonal of the square O x at that moment. And similarly at the end of the 2d, 3d, &c., interval, it will be found at the extremities of the diagonals of the squares next in succession, and as these all lie in one line, O E, 45° inclined both to O P and O p, it appears that in this case the resultant vibration will be rectilinear, and will be performed along the diagonal E G of the square E F G H; and thus it appears that the superposition of two rays of equal intensity, polarized in opposite (*i.e.*, rectangular) planes, results in the production of a ray polarized in a plane 45° inclined to each of the former. Moreover, the square of the diagonal being double that of either side of a square, and the intensity of a ray being measured by *the square* of the vibrational excursion of its ethereal molecules, the intensity of the compound ray will be double that of the components, or, equal to their sum. And, *vice versâ*, any polarized ray may be considered as equivalent to two rays, each of half its intensity, polarized in planes 45° inclined on one side, and on the other of its plane of polarization. It need hardly be observed that if the molecule in starting from O be moving in the direction C A, in virtue of the one vibration, and of c b in virtue of the other,—that is, if it be commencing its first semi-vibration in the one direction, and its second in the other, or again in other words, if the vibrations differ in phase by an exact semi-undulation; all the same reasoning will apply, with this only difference: viz., that the resultant rectilinear vibration will be performed along the other diagonal, H F, of the same square.

(151.) It appears, then, that a change of phase in the vibrations of one of the component rays, of half an undulation, exactly reverses the polarization of the compound ray, and causes its vibration to be performed along the diagonal H F instead of G E. Let us now examine by what sort of gradations the one of these movements passes into the other, when the phase of one of the vibrations C *c* is changed gradually. Suppose, for instance, the vibration *a b* (so, for brevity, we will designate it) to be in advance of the vibration A B by one-twentieth part of a complete undulation, so that at the moment when C starts from C in the direction C A, *c* shall have already got to 1 in the direction *c a*. Then at that moment our molecule O will not be at O but at *y*. After the lapse of one-twentieth more of a period, C will have got to 1 in the direction C A, and *c* to 2 in the direction *c a*, and O, actuated by both movements, will have arrived at *z*, having of course described in the interval a line *y z*, connecting these two extremities of the diagonal of the rectangle *xyz*. And exactly in the same way, at the expiration of the next twentieth of a period, it will be found in *u*, the extremity of the diagonal of the next rectangle—and thus tracing its course step by step through the whole twenty, which constitute a period, we shall see that it will have described a narrow ellipse, having *m n* for its shorter axis, and E G for *the direction of* its longer, and touching the four sides of the square E F G H. If the initial difference of phase be two, three, or four twentieths of the period, it will be seen, by following out the movement in the same way, that more and more open ellipses will come to be described as represented in the figure; and that, when this difference amounts to five twentieths or a quarter undulation, the movement will be circular in the direction *p* P *q* Q, or of the arrow marked thus +. The difference of phases still continuing to increase, this will again degenerate into an ellipse by a continued elongation in the direction H F, and contraction in the direction E G, till it passes at length, after another quarter-undulation of phase-difference, into the straight line H F. The circulation in all, however, being in the same direction, or +. On the other hand, if instead of supposing the vibration *a b* to be initially *in advance* of A B by one-twentieth, we suppose it to be so much *in arrear*, we shall have the same ellipse *n* E *m* described as in the former case, but in the opposite direction, that of the arrow marked —, as will be easily seen by going through the successive steps of our reasoning: and so for all the rest; so that in the case of a circular revolution, the direction of the rotation will be one way or the other, according as the vibration *a c* is a quarter-undulation in advance or in arrear, in respect of phase, of A C.

(152.) This, then, is what is meant by circular and elliptic polarization. It is easy to extend the reasoning above stated to cases in which the component vibrations are of unequal intensity (or extent of excursion), and make other than a right-angle with each other's directions. We have only to suppose our lines A B, *b a*, and their parallels P Q, *p q*, inclined to each other at the angle in question, and of unequal length; to divide them similarly (*i.e.*, in the same proportion) in the points 1, 2, 3, 4, 5—and we shall obtain a set of ellipses, none of which, however, can in either of the cases have its axes equal, or pass into a circle, for this plain reason—that no circle can touch internally all the four sides of any parallelogram except a rhomb.

(153.) Conversely, a ray circularly polarized may be considered as compounded of, and may (by suppressing either of them and letting the other pass, through a tourmaline plate) be resolved into two equal rays, each of half its intensity, polarized at right-angles to each other, and differing in phase by a quarter-undulation. If one of them be in advance of the other by that phase-difference, the rotation will be in one direction—if in arrear, in the other. Elliptic polarization, on the other hand, when it exists, may be recognized by the possibility of resolving the ray so polarized into two oppositely polarized, and either of unequal intensity, or, if equal, differing in phase otherwise than by a quarter-undulation.

(154.) Finally, a ray polarized in any one plane may be regarded as equivalent to two equal rays, circularly polarized in opposite directions of rotation, and having a common zero-point.

(155.) A ray of ordinary light may be considered as a confused assemblage of rays, polarized indifferently in all sorts of planes. It is, therefore, a mixed phænomenon; and to study it in its simplicity, we must in idea break it up into its component elements, and examine their phænomena *per se.* Now it results, from a series of experiments too extensive and refined to be here detailed, and from reasonings upon them which the generality of our readers could hardly be expected to follow, that when a ray, polarized in any plane, undergoes reflexion in a different plane, the reflected portion comes off in all cases more or less elliptically polarized—that is to say, that it consists of, or can be resolved into, two rays, the one polarized in the plane of incidence, the other in a plane at right angles to it—that both these portions have undergone a change of phase at the moment of reflexion, but *not the same for both*, so that arriving at the surface in the same phase, they quit it in different, and therefore constitute by their superposition an elliptically polarized ray. The amount of ellipticity varies, for each reflecting medium (according to the nature of its material) with the angle of incidence at which the reflexion takes place, and also with the inclination of the plane of incidence to that of the primitive polarization of the incident ray. If the reflexion take place on ordinary transparent media of not very high refractive power, as glass, or water, and at the polarizing angle, the degree of ellipticity is so slight that the vibration may be considered as rectilinear, and the reflected ray as completely polarized in the plane of incidence. As the refractive power of the surface increases, the ellipticity impressed is greater, and in some substances, of very high refractive power, such as diamond, and all those bodies which possess what is called the *adamantive lustre* (a consequence of such high refractive power) it is considerable. From such bodies accordingly it is not possible, at any angle of incidence to obtain a reflected ray completely polarized in one plane. And when we come to reflexion from polished metals,* the ellipticity becomes very considerable. In consequence, only a very imperfect polarization of the reflected light in the plane of incidence can be obtained by reflexion from any metalic surface at any angle.

(156.) In all the above enumerated cases, the degree of ellipticity increases *with the reflective power* of the medium of which the reflecting surface is constituted; which itself stands in intimate connexion with the magnitude of the *refractive* index. It might naturally, therefore, be expected to attain its maximum possible amount, or that the ellipse should become a circle in the case of total reflexion. This can only take place, however, when the reflexion is made on the *internal* surface of a transparent medium. This accordingly happens in the case of a beautiful experiment of M. Fresnel, who found that a parallelopiped of glass,† A B C D, fig. 15, being cut and polished, having the acute angles at A and D,

* All metals, even the densest, are in some slight degree transparent, and all have enormously large refractive indices. The transparency of gold is perceptible in gold leaf, which transmits a green light. That of silver is perceptible in the thin films deposited on glass in Liebig's process for silvering mirrors—the transmitted light being bluish.

† The glass used was that known in France as "Verre de St. Gobain."

each 54° 37′, and a ray P Q, polarized in a plane 45° inclined to the plane of the section A B C D intromitted perpendicularly at the face A B, so as to be reflected internally at Q on the side A C, (in which case, the reflexion being at an angle of incidence 54° 37′ was total); and

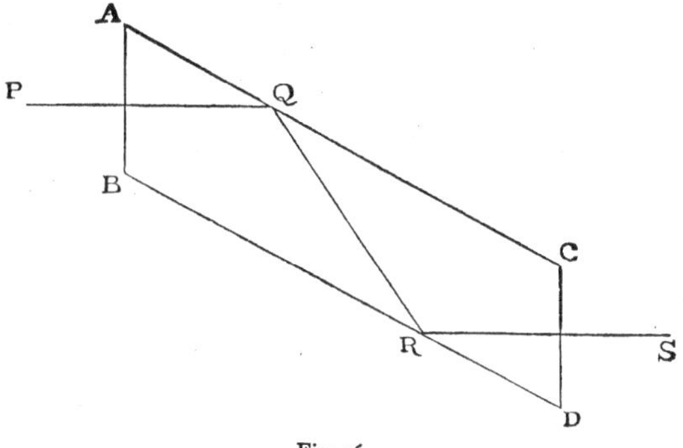

Fig. 16.

again at R, at the same angle, on the opposite side D B, it emerged from the face D C, along the line R S, circularly polarized. In this case, the plane of reflexion making an angle of 45°, with that of original polarization, the reflected ray will consist of two *equal rays*, oppositely polarized; and of these the one in *each* act of reflection has lost, in the other gained, an exact 16th of an undulation, making an 8th *difference* at each reflexion, or a quarter after both; so as to emerge under all the conditions of circular polarization. In consequence, when analysed at its emergence by a tourmaline plate, it is found to undergo no change of brightness on turning the plate in its own plane, whereas the original ray, P Q, would have been wholly extinguished at each quarter revolution.

(157.) Another mode of communicating circular polarization to a ray, is to transmit it at a perpendicular incidence through a parallel plate of any perfectly colourless and transparent doubly refracting crystal, of such thickness, that in the passage through it of the two waves, parallel to its surfaces, into which the incident wave (supposed plane) is divided, (the one conveyed by ordinary refraction, the other by extraordinary, and therefore travelling with different velocities in the crystal,) the one shall have gained or lost, after emergence, exactly a quarter of an undulation on the other. For as the corresponding rays emerge of equal intensity, and oppositely polarized, they here also fulfil all the conditions of circular polarization. If the thickness of the plate be such, that the difference of phases is more or less than an exact quarter (or any number of quarters) of an undulation, the compound ray will be elliptically polarized, and the degree of ellipticity will be determined by the thickness of the plate.

(158.) It may be asked, in what does a ray so circularly polarized differ from an ordinary unpolarized ray, seeing that the latter may always be regarded as compounded of two ordinary rays of half the intensity oppositely polarized? We reply, *in this:* viz., that if again transmitted through another such glass parallelopiped, *similarly situated*, the difference of phase will be doubled. The emergent ray then will consist of two equal rays oppositely polarized (and therefore not interfering), differing in phase by half an undulation, and which therefore (by what we have before shown) compound a single ray polarized in a plane half-way intermediate, or 45° inclined to the original plane of polarization; whereas a ray of ordinary light so transmitted would show no signs of polarization in any one plane more than in any other.

(159.) The most remarkable cases of circular polarization, however, are those which occur when a ray is transmitted along the optic axis of a crystal of quartz, and some few other crystals, as also through certain liquids. The phænomena so exhibited cannot be explained, or even described, however, till we shall have said something

OF THE COLOURS EXHIBITED BY CRYSTALLIZED PLATES ON EXPOSURE TO POLARIZED LIGHT.

(160.) *Uniaxal crystals.*—If a plate cut from a crystal of Iceland spar, so as to have its faces perpendicular to the axis of the primitive rhomboid, be placed close to or very near the eye; and before it a tourmaline plate having its axis vertical, so as to polarize all the light incident upon it in vertical planes passing through the eye; and if any brightly illuminated white surface, such as a white cloud, or a sheet of paper laid in the sunshine, be viewed through it: or if, instead of a tourmaline plate, a "polarizing frame" of glass plates, such as above described, be laid horizontally, and the reflexion of a *clouded* sky be in like manner viewed through the crystal; in the "polarized field" so obtained nothing especially is seen which would lead to a suspicion that the crystal were other than an ordinary piece of glass. But if, in this state of things, between the crystal and the eye, be placed another tourmaline plate, *having its axis horizontal,* a magnificent set of coloured rings will be seen; the exact counterpart of the reflected rings described by Newton (only infinitely more vivid and brilliant), in every respect but these:—First, that they are all divided into four quadrants of coloured light by a dark cross passing through their common centre, and having its arms vertical and horizontal; and, secondly, that the rings themselves are of unequal brightness in different parts of their circumference, being most luminous at the middle points of the quadrants into which the cross divides them, and fading away very gradually on either side of these points, till they cease to be traceable and are lost in the darkness of the cross. On the other hand, if the tourmaline plate between the eye and the crystal (which we shall call the "*analyzing plate,*" or the "analyzer," for a reason which will presently appear) be placed with its axis *vertical,* a series of rings will also be seen: but they are, now, the complementary series—those seen by transmission in the Newtonian experiment; and the cross, instead of black, is now white. Lastly, if the analyzing plate be placed obliquely, both sets of rings will be partially, and, as it were, confusedly, exhibited; the one dislocating the other, in consequence of the brighter annuli of the one set abutting upon the obscurer of the other, the reds on the greens, the purples on the yellows, &c.: the preponderance in light, distinctness, and extent, falling to the share of that set which the position of the analyzer most favours.

(161.) It is manifest that these colours originate in the interference of two series of undulations propagated with different velocities within the crystal, and which therefore must necessarily belong the one to the ordinary, the other to the extraordinary, pencils into which the incident light is divided, which, as before shown, travel with different velocities within its substance. These pencils, however, during their progress through it, are proceeding in different directions (by reason of the double refraction of the medium) and are oppositely polarized—so that, while within the crystal, they cannot interfere. Their interference, then, must be accomplished after their emergence, when their directions have been again reduced to parallelism, and they have been (wholly or partially) brought to a common plane of polarization by the action of the second tourmaline. Let us, therefore, examine how this is brought about. And first, along the vertical arm of the *black* cross, the whole of the incident light being polarized in the plane of a vertical section of the crystal containing its axis, will pass into the ordinary pencil, and none into the extraordinary—so that there will be nothing to interfere with it: and emerging wholly polarized in that plane, will be wholly stopped by the analyzing tourmaline—the result being darkness. But if this tourmaline be turned 90° round in its own plane, it will be wholly transmitted, and the arm of the cross in question will be white. As regards the horizontal arm of the cross, in like manner, the visual ray throughout its whole extent is inclined to the axis in a plane at right angles to that of the primitive polarization. The light, therefore, incident in this plane through the first tourmaline will pass wholly into the extraordinary pencil, and will therefore emerge polarized in a plane *at right angles* to the (now) *horizontal* section of the crystal containing its axis in which its direction lies, *i.e.*, again, in a *vertical* plane, and will be stopped, for the same reason, by the second tourmaline; so that this arm of the cross also will be black in the horizontal, and white in the vertical, position of the analyzer. Let us now consider a ray incident in a plane 45° inclined to the vertical, or in a plane intermediate between the arms of the cross (the axis of the crystal being in all cases supposed held horizontally). The incident ray then will fall on the crystal in a section through its axis 45° inclined to that of its primitive polarization, and will therefore be equally divided between the ordinary and extraordinary pencils. These portions will emerge parallel, and of equal intensity, though differing in phase by such a number of undulations, and parts of an undulation, as the latter, by reason of its greater velocity, has gained on the former. In this state they are both incident on the second tourmaline, having its axis 45° inclined to both their planes of polarization, which therefore will subdivide each of them into two equal portions oppositely polarized, *suppressing or absorbing one, and allowing the other to pass*, and the transmitted portions, being of equal intensity, similarly polarized (viz., both in the plane of the axis of the analyzing plate), and differing in phase, will interfere and give rise to the phænomena of coloration in the manner already sufficiently explained. It remains now to account for the colours being arranged in regular succession in rings round the centre of the black cross (which corresponds to the axis of the crystal). Now the colour developed, or the *order of the tint*, in the series of the Newtonian rings, increases with the difference of phase, and this difference increases with the difference of velocities of the two pencils within the crystal, and with the length of the path traversed with those velocities. Both these increase with the inclination of the visual ray to the axis of the crystal: since along the axis there is no double refraction, which increases gradually from that direction outwards up to a right angle. This, then, explains the progressive increase of colour or order of tint in proceeding from the centre outwards. The circular arrangement is a consequence of the symmetry of the crystalline plate in all directions around its axis; the amount of double refraction being the same at equal obliquities to that line in all directions around it, as also the increase of thickness traversed by rays equally oblique in all directions to the surfaces of the plate. It only now remains to explain how it happens that, in this situation of the analyzing plate (at right angles to the polarizing one), the tints are those of the *reflected*, not of the transmitted series in the Newtonian rings. And the reason is very similar to that by which, in the colours of thin plates, the difference of phase is assumed (justifiably assumed) to commence, not from zero, but from half an undulation. Of the two partial systems of waves that interfere, in the case considered, that which belonged to the ordinary pencil in the crystal passes, as an extraordinary one, through the analyzing plate. Now it is a law, susceptible of demonstration, but which it would lead us too far aside at present to demonstrate—that in the transition from an ordinary to an extraordinary refraction, half an undulation is gained. With the other portion of the interfering pencil, no such transition takes place. Half an undulation then has to be reckoned in addition to the phase-difference due to the simple passage of the two rays through the crystal—just as is the case in the Newtonian reflected rings, and with the same result.

(162.) If we follow out the same chain of reasoning in the case when the analyzing plate is parallel to the polarizing one, the conclusions will be identical up to this last step. But here the cases differ. Neither of the interfering pencils here at its entry into the second tourmaline undergoes extraordinary refraction, and there is accordingly no semi-undulation to be added to the phase-difference. The rings, therefore, will have the characters of the transmitted series of Newton's colours.

(163.) In the generality of uniaxal crystals, the tints of the rings, when the crystal itself is colourless (or as nearly as its colours will allow), follow a succession identical with that of the Newtonian colours of their plates. I have elsewhere called attention, however, to several instances of deviation from this rule, some of which are of so remarkable a nature as to deserve special mention.

The most remarkable is in the case of one variety of the mineral called apophyllite which (from the peculiarity in question) I have proposed to call *Leucocyclite*, in which the rings are almost devoid of colour, being merely a succession of dark and light circles, much more numerous than the coloured ones usually seen, and the more remote of which, from the centre, graduate into feeble shades of purplish and yellowish light. The physical interpretation of this phænomenon is as follows. Since the *colours* originate in the superposition of rings about a common centre, differing in diameter for the several coloured rays throughout the spectrum, (as already explained in Lecture VII.), it follows that in this case, no such difference of diameter, or but a very slight one exists. Now, for crystalline plates so cut, of a given thickness, the apparent diameters of the rings seen are a measure of the doubly refractive energy. The more intense this energy the closer and more compact the system of rings;—for this obvious reason, that the same difference of phases between the ordinary and extraordinary pencils is developed at a less angle of inclination to the axis; and the difference of phases is a direct result of difference of velocities in their internal propagation; and this again, of the doubly refractive energy. Hence we conclude that in the *leucocyclite* all the coloured rays throughout the spectrum undergo equal, or very nearly equal *separation* at a given angle of incidence, by *double refraction;* and that therefore in a doubly refracting prism cut from this substance, the two spectra formed by a sunbeam would be of precisely *equal lengths*, though unequally refracted, or that the highest index of refraction would be accompanied with the least *dispersive power.* I have not made the experiment, but that such would be the case there can be no doubt. In the spectra formed by an Iceland spar prism, the reverse is the case—the higher refractive index corresponding to a much higher dispersive power, and the most refracted spectrum being much longer and much more brilliantly coloured than the least.

(164.) Another highly remarkable example of this kind is found in the mineral called Vesuvian, a uniaxal crystal of a greenish hue, which to a certain degree interferes with the vivid development of its coloured rings. It does not, however, prevent their being well observed—and they present this very singular anomaly, viz., that the system of rings formed by the red rays is considerably *smaller* than those formed by the violet, and in consequence that the order of tints in the rings formed in white light is inverted, so that, of the spectra formed by a prism of this substance, the more refracted ought to be the shorter, and the least coloured. This kind of anomalous action is, however, carried still further in another variety of uniaxal apophylite, in a plate of which perpendicular to the axis, rays of a medium refrangibility *form no rings at all,* so that for such rays the substance *is singly refractive.* Proceeding from this medium refrangibility towards either end of the spectrum, rings are formed, contracting in diameter, as the red or violet end is approached, but most rapidly towards the red. It would not be too much to expect that if a prism could be formed of this mineral (unfortunately very rare), and a bright point illuminated in succession with all the prismatic rays viewed through it, beginning with the red, two images would at first be seen, the one formed by ordinary refraction, fixed, the other gradually approaching it; at a certain stage of the illumination coinciding with it; then crossing to the other side and separating more and more from it as the light verged more to the extreme violet. The experiment, which would be a very beautiful one, is recommended to the attention of those in possession of such crystals which they may not be indisposed to sacrifice.

(165.) *Of the colours developed by circular polarization.*—Quartz, or ordinary *rock crystal* is uniaxal: and when a plate of it of moderate thickness, cut from one of the six-sided prisms in which it usually occurs at right angles to its axis, is examined in the mode above described with a polarizer and analyzing plate, a superb system of coloured rings and black cross is exhibited—but with this peculiarity, that the cross does not come up to the centre, and that the interior rings are blotted out and obliterated by a round patch of coloured light; whose tint, when the tourmalines are at right angles, varies with the thickness of the plate; being white when very thin, and passing, for plates successively increasing in thickness, through all the series of tints of Newton's transmitted rings. Keeping to one plate, the tint also varies on turning round the analyzing plate in its own plane, and with this very extraordinary peculiarity, viz., that while in some crystals a certain succession of colours is observed, on turning it from right to left; in plates of the same thickness cut from other crystals the same succession is seen on turning it from left to right. Yet more singular, is the fact that this inversion—this right-and-left-handedness in the succession of tints, corresponds to, and is predictable beforehand from, the appearance of certain small obliquely posited facets on the crystal previous to polishing, which lean unsymmetrically in some crystals to the right, in others to the left hand of the axis held up straight before the eye. In all other respects the crystals are identical.* A similar right-and-left-handedness in the external form of their crystals, accompanied with the very same optical phænomena, has been remarked by M. Pasteur in the salts called *para-tartrates* and their crystallized acid.

(166.) The account given by the undulatory theory of these phænomena is this. Quartz (to adhere to our first, chosen instance) is uniaxal, but it differs from Iceland spar and others of that class in a most essential point first noticed by Mr Airy, viz.: that the sphere and spheroid representing the simultaneous surfaces of the ordinary and extraordinary waves propagated within them, though having a common axis, do not touch each other internally. Hence, in the direction of that axis, though there is, *at a perpendicular incidence*, no double refraction, there *is* a difference of velocity in the two rays. Now the theory at present adopted is, that owing to some peculiarity at present not understood; when a polarized

* Amethyst consists of thin alternate layers of right-handed and left-handed quartz superposed, parallel to their axes.

ray (which may always be considered as compounded of two circularly-polarized ones of opposite characters as already stated, *i.e.*, in which the particles of the ether circulate in opposite directions) is incident on a quartz plate, in this manner; the crystal operates an analysis of the ray and *resolves it into* two such rays circularly polarized; which it propagates *as such*, the one as an ordinary, the other as an extraordinary one. On their emergence at the opposite face of the plate they recompound a plane-polarized ray; but, having gained or lost on one another, by reason of their difference of velocity in their passage through it, a number of revolutions or parts of a revolution proportional to the thickness of the plate, the two circular rays at the instant of their reunion have no longer a *common zero-point* as at their entry: and from this it may be demonstrated* that the plane of polarization of the recomposed will not be coincident with that of the incident ray, but will have been turned round, *in the direction of the rotation of the ray which travels fastest within the quartz*, through an angle also proportional to the thickness of the plate. As the angle of displacement, moreover, differs for the differently coloured rays of the spectrum; the effect will be that, when passed through an analyzing tourmaline the different colours will be differently absorbed, and the result will be the production of a compound tint in the beam finally delivered into the eye, the colour of which will vary with the rotation of that plate in its own plane, as observed.

* Our necessary limits forbid us to give the steps of the demonstration, which, however, are very obvious.

(167.) It is not in crystallized bodies only that this singular effect is produced. Strange as it may seem that a colourless, transparent, and perfectly homogeneous *fluid* should deviate the plane of polarization of a ray passing perpendicularly through it at all; still stranger that it should do so constantly in one direction for the same fluid, but in opposite directions for different fluids; strangest of all, that even vapours should be found possessing the same property: such is the case. Thus, oil of turpentine and its vapour turn the plane of polarization to the right hand, solution of sugar to the left, and so for a variety of other substances.* This property has been made the basis of an elegant instrument called the saccharometer, by which the quantity of sugar contained in a given solution is ascertained by simple inspection of the tint so produced.

(168.) Struck by the fact, apparently so singular, of a "right-and-left-handedness" inherent as it were in the molecules of material bodies—by the correlative fact of such a tendency, or so to speak idiosyncrasy, manifesting itself in the forms of crystals—and again, in quite a different field of scientific research, in the action of an electrified cylindrical wire on a magnetized needle placed parallel to its direction, (which turns the north end of the needle to the right or to the left according to the direction of the current *along* the wire): it early occurred to the writer of these pages that it was

* Mr Jellett, of Trinity College, Dublin, has, I am informed, recently discovered a liquid which is *right-handed* for one end of the spectrum, but *left-handed* for the other!

scarcely possible such singularities should stand in no natural connexion. Between two of the cases adduced the connexion had been proved by himself. It remained to enquire whether the third could be brought into obvious relation to the other two. Accordingly on the 14th of March 1823, having prepared a long spiral coil of copper wire enclosed in an earthenware tube, furnished with a polarizing reflector at one end and an analyzer at the other; by the kindness of the late Mr Pepys, he was permitted to bring the coil into connexion with the great magnetic combination of the London Institution, consisting of one enormous couple, expressly arranged for producing the greatest possible magnetic effect. His expectation was that light would appear in the dark polarized field on making the contact, and be maintained during its continuance. The experiment, however, proved unsuccessful. No *direct* action upon *light* could so be made manifest. At a later period, however (1845), by introducing into a similar coil a certain highly refractive glass consisting chiefly or wholly of borate of lead, as well as a variety of other solids and liquids (water among others), Professor Faraday succeeded in communicating, temporarily, and during the continuance of the passage of the current, the property in question to them.

(169.) *Biaxal Crystals.*—By far the greater number of crystallized substances do not present that *single symmetry* (symmetry on all sides of a single central line or axis), which we have spoken of as indicative of a single axis of double refraction, and of a spherical propagation of the ordinary, and spheroidal of the extraordinary ray, within them. In all or nearly all these, two lines inclined to one another at an angle greater or less according to the nature of the substance can always be found (either by a careful examination of the crystal in polarized light through the faces of its natural form, or by cutting and artificially polishing plates of it), which possess the properties of such axes;—along which, that is to say, refraction is single for a ray passing either way out of the crystal; and in which when examined in polarized light with an analyzing plate between the eye and the crystal, coloured rings are seen. The simplest and readiest instance of a crystal of this kind is furnished by a sheet of ordinary mica, such as may easily be procured in large sheets. If a sheet of this be held before the eye in a polarized field perpendicularly (an analyzing plate being interposed) and turned round in its own plane, two portions will be found at right angles to each other, in which the polarization of the incident light is not disturbed, and the field remains dark. Of the two planes perpendicular to the plate in which the plane of polarization cuts it in these two positions, one is the "principal section" of the plate, and contains its "optic axes." These may be brought into sight by holding the eye (armed with the analyzer) quite close to the mica, and inclining the latter, either forward or backward in one of these two section-planes so as to make an angle of about 35° with the visual ray on either side of the perpendicular. In either situation a set of coloured rings will be seen, not circular, but of an oval form, about a common centre, and intersected, not by the two arms of a black cross as in Iceland spar, but by one vertical dark bar cutting centrally across them. This dark bar is converted to a white one, and the colours of all the rings changed to their complementary ones, by turning the analyzing plate through 90° in its plane.

(170.) In mica, the angular separation of the optic axes is too large to allow both these sets of rings to be seen at once, so as to examine the nature of their mutual connexion. In nitre however, in which it is only about 5° (within the crystal), this may be very conveniently done, by cutting from the clear transparent portion of a large hexangular-prismatic crystal (such as may always be found in searching over a lot of the ordinary commercial saltpetre) a plate about a quarter of an inch thick, perpendicular to the axis of the prism, and polishing its faces. If this be placed between two crossed tourmalines, and held up against the light, the normal phænomenon of the biaxal rings will be seen in its utmost perfection, as in fig. 17, the upright and horizontal lines in which indicate broad brushes as it were of shadow, cutting across the system of ovals, and breaking them up into four similar quadrants. If, retaining the tourmalines in the same position, the nitre plate be turned round in its own plane, this cross breaks up into two curved arcs, as represented in fig. 18, corresponding to a movement through a quarter of a right angle,—then, as in fig. 19, corresponding to 45° of change, and so on till after a quarter of a revolution the original appearance of fig. 17 is restored. If the tourmalines instead of being crossed, are laid parallel, the forms of the ovals are the same, but the colours complementary, and the cross and curved branches white.

(171.) The forms of these curves are governed by a

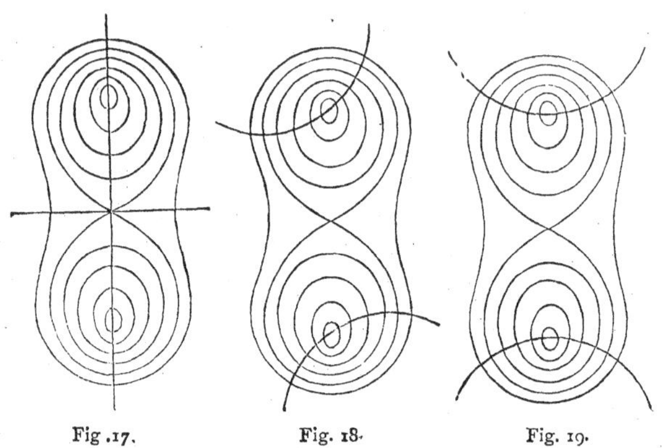

Fig. 17. Fig. 18. Fig. 19.

very simple and elegant general law, common to all biaxal crystals, and applicable to every angular separation of the axis: and when this separation is small, as in the case before us, they may be regarded as "lemniscates," of which the property is this, that for every point in the circumference of each oval the *product* (not as in an ellipse the *sum*) of two lines drawn to the two centres or foci, is invariable; and for successive ovals proceeding outwards from either focus, these products increase in regular arithmetical progression. When the two foci coincide, that is to say, when the two axes of the crystal coalesce, and it becomes uniaxal, the ovals pass into circles, and we fall back upon the circular rings and cross proper to that class of bodies.

(172.) Neglecting the bending which the rays undergo at their emergence from the posterior surface of the crystal, or conceiving the eye as immersed within its substance, it is evident that when looking in the direction of either of the foci of the ovals, the visual ray will be directed along one of two axes, or *lines of no double refraction;* while if looking towards any point in the circumference of any one of the ovals, the visual ray will traverse the crystal in such a direction that an ordinary and extraordinary ray following that path shall gain or lose on each other so many semi-undulations, or parts of one, as shall correspond to the tint developed in that direction; and that, therefore, in all the directions marked out by the circumference of each individual oval, the tints being the same, the phase-difference, *and therefore* the difference of velocities of the interfering rays, *and therefore again*, the amount of double refraction in that direction is the same. The forms of these ovals, therefore, stand in immediate and intimate connexion with the law of double refraction in such crystals, and with the forms of the two wave surfaces belonging to the ordinary and extraordinary rays. The theory of these wave-surfaces belongs, however, to a higher department of geometry than we could hope to make intelligible in these pages. Suffice it to say that as delivered by M. Fresnel and his followers it explains all the facts in the most complete and satisfactory manner, and has even led to the prediction, antecedent to observation, of some phænomena so apparently paradoxical as to stand in seeming contradiction with all previous optical experience; and which any one, antecedent to their verification by trial, would have pronounced impossible.*

(173.) One highly important conclusion from this theory must, however, be noticed. The directions within the crystal of the two axes of double refraction or the "optic axes" stand in no abstract *geometrical* relation to those of the angles and edges of its "primitive form," or to its axes of symmetry. They are resultant lines determined by the law of elasticity of the luminiferous ether within its substance as related to its crystalline form, and *to the wave-length of the particular coloured ray transmitted.* They are not, therefore, the same for all the coloured rays. In the generality of biaxal crystals, the difference of their situations and of the angle between the two, is but small: but in some, as in the salt called Rochelle salt (tartrate of soda and potash), it is very great, amounting to at least 10°, by which the direction within the crystal of either axis for the extreme red rays differs from that for the extreme violet.† In this salt the variation in position of the optic axes progresses pretty uniformly in passing from a red to a violet illumination. In Carbonate of lead, on the other hand, it varies slowly in

* This alludes to the phænomena of what is called conical refraction, pointed out by the late Sir Wm. R. Hamilton, as a necessary consequence of Fresnel's theory, and demonstrated to exist as a matter of fact, subsequently, by Dr Lloyd.

† See a paper by the author of these pages in *Phil. Trans.*, 1820, "On the action of crystallized bodies on homogeneous light," where the singular phænomena to which this gives rise are fully described.

passing from red to green, but with increasing and finally with extreme rapidity in the passage thence to violet.

(174.) It is time, however, to bring this long Lecture to a conclusion. To describe the variety of splendid and singular phænomena, developed in every department of physical enquiry by the use of polarized light, not one of which has hitherto afforded any, the smallest, ground for doubt as to the applicability of the undulatory theory to its complete explanation, would require volumes. We would gladly have said something of the magnificent phænomena, exhibited by "macled" crystals,* and by unannealed, or compressed glass; of the changes produced by change of temperature on the optical relations of bodies, and of the calorific and chemical rays of the spectrum; but our limits forbid it. Suffice it to add, that what the telescope and the microscope effect for us in the discovery of outward and visible form, the properties of light, and especially of polarized light, effect in subjecting to our intellectual vision the intimate structure of material bodies. Within the compass of the smallest visible atom they open out a world of wonders—a universe *sui generis*, and for each atom of a different material a different one—all, however, related and bound together in one vast harmony.

* These may find a place elsewhere. The phænomena alluded to have not, so far as I am aware, been hitherto described.

LECTURE IX.

ON SENSORIAL VISION.*

FROM what I have understood respecting the objects of this Institution, as one equally addressing itself to the cultivation of Philosophy and Literature, I am led to believe that the subject which I propose to bring before it, in compliance with an invitation which I feel to be both honourable and gratifying in no common degree, is one not altogether foreign to them; inasmuch as the history of vision has both a strictly scientific and a more abstract and philosophical bearing; the one referring itself to material, the other to mental science. The one regards only the means and adaptations by which *we see*, and the other refers to the action of the mind itself *in seeing*, that is, in *interpreting* the impressions produced on our visual

* This lecture was delivered before the Philosophical and Literary Society of Leeds, on the 30th September 1858, at the request of that society.

organs. It is to this latter division of the subject that I shall chiefly address myself, while taking the opportunity thus kindly afforded me of putting on record certain visual phenomena which I have from time to time noticed, belonging to that obscure class of impressions which may be termed Sensorial Vision—by which I mean visual sensations or impressions bearing a certain considerable resemblance to those of natural or *retinal* vision, but which differ from these in the very marked particular of arising when the eyes are closed and in complete darkness.

(2.) Few persons, I suppose, are ignorant, as a matter of personal experience, of the sort of appearances known by the name of Ocular Spectra, which are produced by the impression of a strong light on the retina of the eye, and which continue to force themselves on the attention, sometimes in a very pertinacious and disagreeable way for some time afterwards, when the eyes are closed. In one lamentable instance, that of an eminent Belgian Philosopher, they have caused actual loss of sight; and in that of Sir Isaac Newton, their obstinate recurrence is said to have deprived him of sleep for several days and nights successively, and to have driven him to the verge of distraction. These are cases when the stimulus of light has been pushed to the extreme; but when moderate and regulated, these spectra admit of being studied: and the laws of their production—the singular and beautiful phases they pass through—their periodical extinction and renewal (which extend over a very considerable interval of time from their first production), the orderly

2 C

recurrence of the colours they assume, which are peculiarly rich and various—the singular effects of gentle pressure on the eye, and partial light admitted through the eyelids in modifying them or in renewing them when extinct—all these offer a subject of much attraction and interest. A very interesting memoir on them has been, within these few years, communicated to the Royal Society, by Dr Scoresby; but the subject is far from being exhausted, and it is to the habit of attention to such sensorial impressions, fostered by frequently watching the development of these spectra under a variety of circumstances in my own case, that I attribute my having been led to notice that other class of phænomena of which I shall presently speak, and which from their inconspicuousness, I suppose, escape the notice of most people.

(3.) The production of Ocular Spectra refers itself, I presume, to what I have described as the purely physical branch of the general subject of vision. Their seat, it can hardly be doubted, is the retina itself,* and their production is in all probability, part and parcel of that photographic process by which light chemically affects the retinal structure, and of the gradual restoration of that structure to its normal state of sensitiveness by the fading out of the picture impressed. Cases are not wanting in artificial photography where an impression made

* In speaking of the retina, I would not be understood to express any opinion on the disputed question whether the *retina* anatomically so called or the choroid coat of the eye be really the seat of vision.

on sensitive paper dies out, and can be replaced by another without the renewed application of any chemical agent.

(4.) Thus considered, ocular spectra are quite as much entitled to be considered as things *actually seen*, as the retinal pictures of which they are the successors, or rather remnants. It is quite otherwise with that other class of visual impressions to which I now refer, and which differ altogether from ocular spectra, not only in being for the most part (though not always) much less vivid and much more dreamy (if I may use the term without casting a doubt on their reality *as facts*), but also in having no reference or resemblance to any objects recently seen, or even recently thought of. Of course, when I speak of their reality as facts, I do so on the ground of their admitting of being watched and studied with the same sort of wide-awake attention which might be given to any faint and fugitively-presented real object: though it is no more possible to describe them accurately, much less to draw them, than it would be to do so in the case of objects dimly seen in the dusk of evening, and capriciously appearing and disappearing. But this does not preclude their being observed and described, *pro tanto*, in general terms.

(5.) I fancy it is no very uncommon thing for persons in the dark, and with their eyes closed, to see, or seem to see, faces or landscapes. I believe I am as little visionary as most people, but the former case very frequently happens to myself. The faces present themselves involuntarily, are always shadowy and indistinct in outline

—for the most part unpleasing, though not hideous; expressive of no violent emotions, and succeeding one another at short intervals of time, as if melting into each other. Sometimes ten or a dozen appear in succession, and have always, on each separate occasion, something of a general resemblance of expression or some peculiarity of feature common to all, though very various in individual aspect and physiognomy. Landscapes present themselves much more rarely but more distinctly, and on the few occasions I remember, have been highly picturesque and pleasing, with a certain but very limited power of varying them by an effort of the will, which is not the case with the other sort of impressions. Of course I now speak of waking impressions, in health, and under no kind of excitement. When the two latter conditions are absent, numerous instances are on record of both voluntary and involuntary impressions of this kind, and singular as some of the facts related may appear, I am quite prepared, from my own experience on two several occasions, to receive such accounts with much indulgence.

(6.) A great many years ago, when recovering from fever, my chief amusement for two or three days consisted in the exercise of a power of calling up representations both of scenes and persons, which appeared with almost the distinctness of reality. One of these scenes I perfectly recollect. A crowd was assembled round a hole in the ice, into which a youth had fallen. His mother was standing in agony on the brink, and there were the floating fragments and something of a shadowy form under the blue transparent ice.

In this case there was, of course, the excitability of nerve connected with the remains of bodily disorder. On the other occasion to which I allude, I had been witnessing the demolition of a structure familiar to me from childhood, and with which many interesting associations were connected: a demolition not unattended with danger to the workmen employed, about whom I had felt very uncomfortable. It happened to me at the approach of evening, while, however, there was yet pretty good light, to pass near the place where the day before it had stood; the path I had to follow leading beside it. Great was my amazement to see it as if still standing—projected against the dull sky. Being perfectly aware that it was a mere nervous impression, I walked on, keeping my eyes directed to it, and the perspective of the form and disposition of the parts appeared to change with the change in the point of view as they would have done if real. I ought to add, that nothing of the kind had ever occurred to me before, or has occurred since. On this occasion, no doubt, the daily habit of seeing the same object from the same point of view for years would naturally give great efficacy to the associative principle, and the fact can only be regarded as an exemplification of a physiological process which I shall presently have occasion to speak of more particularly.

(7.) But it is not to phænomena of this kind that I am about specially to direct your attention. The human features have nothing abstract in their forms, and they are so intimately connected with our mental impressions that the associative principle may very easily find, in casual and irregular patches of unequal darkness, caused by slight local pressure on the retina, the physiognomic exponent of our mental state. Even landscape scenery, to one habitually moved by the aspects of nature in association with feeling, may be considered as in the same predicament. There is nothing definite or structural in its forms, which are arbitrary to any extent, and composed of parts having no regular or symmetrical relations. It is perfectly conceivable that the imagination may interpret forms, in themselves indefinite, as the conventional expressions of realities not limited to precise rules of form. We all know how easy it is to imagine faces in casual blots, or to see pictures in the fire.

(8.) But no such explanation applies to the class of phænomena now in question, which consist in the involuntary production of visual impressions, into which geometrical regularity of form enters as the leading character, and that, under circumstances which altogether preclude any explanation drawn from a possible regularity of structure in the retina or the optic nerve.

(9.) I was sitting one morning very quietly at my breakfast-table doing nothing, and thinking of nothing, when I was startled by a singular shadowy appearance at the outside corner of the field of vision of the left eye. It gradually advanced into the field of view, and then appeared to be a pattern in straight-lined-angular forms, very much in general aspect like the drawing of a fortification, with salient and re-entering angles, bastions, and ravelins, with some suspicion of faint lines of colour between the dark lines. The impression was very strong:

equally so with the eyes open or closed, and it appeared to advance slowly from out of the corner till it spread all over the visual area, and passed across to the right side, where it disappeared. I cannot say how long it lasted, but it must have been a minute or two. I was a little alarmed, looking on it as the precursor of some disorder of the eyes, but no ill consequence followed. Several years afterwards the same thing again occurred, and I recognized, not indeed the same precise form, but the same general character—the fortification outline, the dark and bright lines, and the steady progressive advance from left to right. I have mentioned this to several persons, but have only met with one to whom it has occurred. This was a lady of my acquaintance, who assured me that she had often experienced a similar affection, and that it was always followed by a violent headache, which was not the case with me. In this case the regularity of the pattern was not great, but the lines were quite straight and the angles sharp and well defined. Had it remained stationary, it might be assumed that the retina had a structure corresponding to the figure; and that some undue pressure might render that structure visible. But such an hypothesis is precluded by the gradual transit of the lines over every part of the visual area.

(10.) I come now to cases of perfect symmetry, and geometrical regularity. The most ordinary class of patterns of this sort I find to be formed only in darkness, and if the darkness be complete, equally with open as with closed eyes. The forms are not modified by slight pressure, but their degree of visibility is much and capriciously varied by that cause. They are very frequent. In the great majority of instances the pattern presented is that of a lattice work; the larger axes of the rhombs being vertical. Sometimes, however, the larger axes are horizontal. The lines are sometimes dark on a lighter ground, and sometimes the reverse. Occasionally at their intersections appears a small, close, and apparently complex piece of pattern work, but always too indistinctly seen to be well made out. The lattice pattern if constant, and if always upright, might be explained by the habit of looking fixedly at a lattice window, with a view to noting the order of succession of colours in the Ocular Spectra, which this mode of viewing them shows finely. Occasionally, however, the latticed pattern is replaced by a rectangular one, and within the rectangles occurs in some cases a filling in of a smaller lattice pattern, or of a sort of lozenge of fillagree work, of which it is impossible to seize the precise form, but which is evidently the same in all the rectangles. Occasionally too, but much more rarely, complex and coloured patterns like those of a carpet appear, but not of any carpet remembered or lately seen, and in the two or three instances when this has been the case, the pattern has not remained constant, but has kept changing from instant to instant, hardly giving time to apprehend its symmetry and regularity before being replaced by another; that other, however, not being a sudden transition to something totally different, but rather a variation on the former.

(11.) Hitherto I have mentioned only rectilinear forms.

I come now to circular ones. Having had to submit to a surgical operation, I was put under the blessed influence of chloroform. The indication by which I knew when it had taken effect consisted in a kind of dazzle in the eyes, immediately followed by the appearance of a very beautiful and perfectly regular and symmetrical "Turks-cap" pattern, formed by the mutual intersection of a great number of circles outside of, and tangent to, a central one. It lasted long enough for me steadily to contemplate it so as to seize the full impression of its perfect regularity, and to be aware of its consisting of exceedingly delicate lines; which seemed, however, to be not single but close assemblages of coloured lines not unlike the delicate coloured fringes formed along the shadows of objects by very minute pencils of light. The whole exhibition lasted, so far as I could judge, hardly more than a few seconds, and I should observe that I never lost my consciousness of being awake, and in full possession of my mind, though quite insensible to what was going on. I spoke, but the words I am told I uttered had no relation to what I know I meant to say.

(12.) After a considerable interval of time it became necessary to undergo another operation, which was also performed under chloroform, but this time the dose was less powerful, or differently administered. Again the "Turks-cap" pattern presented itself on the first impression, which I watched with much curiosity, but it did not seem quite complete, nor was it identical with the former. In the intersections of the circles with each other, I could perceive small lozenge-shaped forms or minute patterns, but not clearly enough to make them well out. On both these occasions the patterns were far more lively and conspicuous than the dim and shadowy forms before spoken of, and probably belong to quite a different class of phænomena.

(13.) Since that time circular forms have presented themselves spontaneously, of the shadowy and obscure class, on three occasions, one of them quite recently. On the first of these, circular were combined with straight lines forming a series of semicircular arches, supported by, or rather prolonged beneath into, tall slender vertical columns, the whole like small wirework; mere lines, and bright on a dark ground; while another series of similar arches and uprights darker than the general ground appeared, intersecting the former so as to have the dark uprights just intermediate between the bright ones of the first set. On the second occasion the pattern consisted of a very slender and delicate circular hoop, surrounded with a set of other circles of the same size, exterior tangents to the central circle and to each other. On the third, the whole visual area was covered with separate circles, each having within it a four-sided pattern of concave circular arcs. All these phænomena were, however, much fainter than the chloroform exhibitions, and of the order of the lattice patterns.

(14.) Now the question at once presents itself—What *are* these Geometrical Spectra? and how, and in what department of the bodily or mental economy do they originate? They are evidently not dreams. The mind is not dormant, but active and conscious of the direction of its thoughts; while these things obtrude themselves on notice, and, by calling attention to them, *direct* the train of thought into a channel it would not have taken of itself. Retinal impressions they can hardly be, for what is to determine the incidence of pressure or the arrival of vibrations from without upon a geometrically devised pattern on the retinal surface, rather than on its general ground. The effect of some cause in the nature of pressure I on one occasion experienced, and it manifested itself quite differently, viz:—as an Ocular Spectrum, consisting in a round, deep purple, feebly luminous spot, dying gradually away into darkness at the borders. It was not exactly in the middle of the visual area, and was caused by no external light: for it was perceived one morning immediately on waking in the morning twilight, and with the face shaded from a direct view of the window.

(15.) It is quite clear that a regular geometrical pattern cannot be suggested to the imagination by forms having no regularity, however presented to it: so that the explanation which in the other instances adduced might have a certain plausibility, breaks down in these cases. It may be said that the activity of the mind, which in ordinary vision is excited by the stimulus of impressions transmitted along the optic nerve, may in certain circumstances take the initiative, and propagate along the nerve a stimulus, which, being conveyed to the retina, may produce on it an impression analagous to that which it receives from light, only feebler, and which, once produced, propagates by a reflex action the sensation of visible form to the sensorium. Still, even granting that such reflex action is possible, and the retina is so impressed, the question remains—Where does the pattern itself or *its prototype in the intellect* originate? certainly not in any action *consciously* exerted by the mind, for both the particular pattern to be formed and the time of its appearance are not merely beyond our will and control, but beyond our knowledge. If it be true that the conception of a regular geometrical pattern implies the exercise of thought and intelligence, it would almost seem that in such cases as those above adduced we have evidence of a *thought*, an intelligence, working within our own organization distinct from that of our own personality. Perhaps it may be suggested that there is a kaleidoscopic power in the sensorium to form regular patterns by the symmetrical combination of casual elements, and most assuredly wonders may be worked in this way. But the question still recurs in another form: "How is it that we are utterly unconscious of the possession of such a power; utterly unable voluntarily to exert it; and only aware of its being exerted at times, and in a manner we have absolutely no part in except as spectators of the exhibition of its results?"

(16.) But again, it may be urged that the particular geometrical forms presented are familiar ones, and are not created or invented *pro re nata*, but simply old ones reproduced—their reproduction being an act, not of invention, but of memory. But against this view of the matter there appears to me to exist an insuperable objection. Memory does not produce its effect by creating before the eyes a visible picture of the object remembered. When Hamlet says, "Methinks I see my father," we all know that the expression is a purely figurative one, and have no need to be told, as in his reply to Horatio's "Where? my lord," (a question perfectly natural to one who had just seen his ghost and knew not but that it might still be present), that it is to the "mind's eye," a merely figurative and metaphorical eye, and not to that of the body, that the expression applies. The act of *reminiscence* is a *conscious* and a mental act, and if, under the influence of powerful excitement and strong associations it ever results in the production of a visible picture by the sort of reflex action I have described, it must precede such formation—or any how not be itself called up by the picture of its own creation. Of such cases whenever they occur (and I have related what may be considered a case in point) the same account is to be given as in that of certain eminent painters, who are said to have declared that they *see* upon the paper or the canvas the forms they are about to delineate—a quasi-image being formed on the retina by the sympathy of the nerve with the brain, and its impression delivered back to the sensorium as that of a reality.

(17.) I ought perhaps to apologize for saying so much about myself and my personal experiences in this line, but the nature of the subject is such as to render this inevitable; and it is one which can only be elucidated by the individual putting on record his own personal contribution to the stock of facts accumulating. And having gone so far in this direction, I may perhaps be borne with if I add one or two more observations of a similar personal nature, which, though not bearing on the subject hitherto spoken of, seem to me not without some interest as contributions to that mass of unaccountable or difficultly explicable facts with which the history of vision teems so abundantly.

(18.) The first of these is one of constant occurrence to myself in railway travelling. When looking out on a sloping bank, the train going rapidly, if the sight be directed fixedly out in one direction, all near objects—stones, grass tufts, &c.,—are of course seen as if drawn out into horizontal lines. Now what I constantly perceive is the appearance of slender obscure lines like dimly seen dark wires at regular intervals asunder, crossing those linear streaky images nearly at right angles, and which always seem not to stand vertically up and down, but as if they reclined backwards on the slope of the bank. I find it best to let the eyes take their own focus without endeavouring to adjust them to any object.

(19.) It is generally taken for granted that to see any object whatever, the best way is to look straight at it, and get its image impressed on the centre of the retina. This is certainly, however, not the case with a single bright luminous point, if no brighter than a star of the third or fourth magnitude, as any body may convince himself by trying the experiment the first clear night. When two such stars of equal magnitude, within a degree or two of each other, are looked at, nothing is easier than to make either of them disappear as if blotted out from the sky, by looking full and fixedly *at* it, while the other remains conspicuously visible. In this way I find stars of the second magnitude considerably enfeebled, though they cannot be made wholly to disappear. Those of the first are but little affected. I have found many persons incredulous on their first hearing of this fact, who yet have satisfied themselves by trial of its reality. I at one time believed that this comparative insensibility of the centre of the retina arose from the greater wear and tear consequent on directing the attention continually to it, and habitually directing it to any more conspicuous object, but I find that the same thing happens to very young persons to quite as great an extent—in whom, of course, this cause of deterioration cannot have gone so far as in adults. There is reason to believe moreover, that this comparative insensibility of the middle part of the retina to faint impressions extends over a pretty considerable area, for I find that in a room but feebly lighted, and with the back to the light, it is possible by long looking fixedly *in the direction* of an object of considerable angular diameter, gradually to lose sight of it, and at length entirely cease to see it—and then, by an effort of the will, accompanied with some kind of organic act in the eye itself, which I know by sensation, but am unable to describe in words, but which is not the action of adjusting the focus, it is at once realized to sight, without any alteration of the direction of the optic axis, or any motion given to the head or person. It is an experiment which will not always succeed, and requires a peculiar adjustment of the light, and of the comparative illumination of the objects and the ground on which it is seen projected, and perhaps also a peculiar state of nerve; but when it does succeed, the effect is exceedingly singular and anomalous.

(20.) It would lead me into too great a length of detail, and I may also add, into a labyrinth of metaphysical considerations, out of which I should find some difficulty of getting disentangled, if I were to go into a discussion on those points of connexion between our mental and our bodily organization which these facts seem to suggest. There is a very curious chapter in Stuart Mill's *Treatise on Logic*, devoted to the question whether we are quite sure that every event *has a cause.* He decides it, as every reasonable man must do, in favour of the universality of the proposition, but he is compelled to admit, as every one who considers it closely must, I think, equally do, that the phænomena of the human will stand in a very peculiar relation to that question; and that granting volition to be a *cause* of action, and granting the entire *freedom* of our will and its complete *independence* to choose when a choice of lines of action is brought before us, there is still the question behind—What determines the will? To this question an answer must be found which will leave man a moral and responsible agent. To choose the right and to avoid the wrong, *as such*, must be left in his power, and a freedom and independence of choice as between these two grand lines of action must be left him, if we would not reduce him to a machine. So far then, and to this extent, I do not see how it is possible not to recognize an original *causation*, or at least one which it is morally, intellectually, and logically impossible *for us* to find an antecedent for by any power of merely human inquiry. But still there arises this other and further question—What determines the will in cases where a variety of modes of action exist; all, so far as we can see, equally open to choice? Mr Mill here refers us to the associative principle; and refers the moral position of the individual to the education or early discipline of this associating principle, by which it may be habituated to suggest right and virtuous courses of action among the many possible ones more readily, more powerfully, and more *suggestively* (if one may tautologize so far) than those of a contrary nature. It is very evident that with the greatest rectitude of intention, if a course of action the most conducive to the interests of good do not suggest itself, or be not suggested from without, the course actually adopted may be one less so. It is then to the suggestive principle, whatever that be, and however it may act, that we must look for much that is determinant and decisive of our volition when carried out into action, even when the choice has been made between right and wrong in the abstract; and the "way in which thoughts come into our minds" is part and parcel of the nature and mode of action of that principle—if it be not merely another form of words for the same thing. Of course this is a subject so obscure and so mysterious, that it is quite out of the question to pretend to raise any theory—state any doctrine—or say anything in the least degree satisfactory, or on which any two opinions would agree, about it. Still it strikes me as not by any means devoid of interest, to contemplate cases where, in a matter so entirely abstract, so completely devoid of any moral or emotional bearing as the production of a geometrical figure, we, as it were, seize upon that principle in the very act, and in the performance of its office.

(21.) I will not, however, pursue this further, lest I get bewildered myself and bewilder all who listen to me, and it only remains for me to thank you for your patience in listening to me so far, and to apologize once more for dwelling upon my personal impressions. There is one thing more I would add—which is to congratulate this town of Leeds on the existence within it of a society of such a nature as that here assembled, and numbering among its members so many gentlemen of large and liberal views, of extensive information and high acquirements, both scientific and literary—equally the ornaments of this their city, and of our common country.

LECTURE X.

THE YARD, THE PENDULUM, AND THE METRE.

CONSIDERED IN REFERENCE TO THE CHOICE OF A STANDARD OF LENGTH.*

THE attention of the public has of late been strongly drawn to the subject of a proposed alteration of our national system of weights and measures, by the attempt made during the last session of Parliament to carry through a bill, having for its object the abolition of our existing system in its entirety, and the introduction, in its place, of what is known as the "French Metrical System." The bill, it is true, was withdrawn after passing the second reading (by which the House, as is usually supposed, "*affirmed*

* This Lecture or Essay was communicated to the Leeds Astronomical Society, and read at a meeting of that Society on October 27, 1863.

the principle of the measure"), and it may therefore be reasonably presumed that it will be brought forward again in the next session, in the same or a modified form. As the discussion it received in the House seemed to be in no respect commensurate with the immense importance and sweeping nature of the change proposed, and with the exception of one or two rather cursory notices in *The Times*, excited a marvellously small amount of public interest pending its progress; it will not be amiss if, being called upon by the committee of the Leeds Astronomical Society for an exposition of some point of general interest in the form of a Lecture or Essay, to be read at one of their Evening Meetings, I select this for its subject; and endeavour to place before you the several conditions which any standard or typical unit of length which shall be assumed as the basis of a system of measures and weights intended to be national, and which may justly claim to be universal, ought to fulfil; and to compare with these conditions, in order to see how far they are fulfilled in fact, both our actual standard, the French metre now in use, and the length of the pendulum, which has been more than once proposed as a natural unit of length. And this I will endeavour to do in as elementary and familiar a way as shall be consistent with perfect correctness. Those of the present audience who are not already familiar with the subject will thus be better enabled to form an opinion as to the desirableness of the change actually proposed, or of any legislative change in our existing standard, and in our system of measures, weights, and coinage generally. And to such it will not be amiss to observe in the outset that, the subject being an exceedingly delicate and refined one, they must not be surprised at seeing very minute quantities and very small fractions treated as matters of much greater importance than they may have been accustomed to regard them.

(2.) The general subject of a national system of weights and measures, be it observed, divides itself into two very distinct and separate points of inquiry, viz.: first, What is intrinsically the best and most available unit of linear measure to adopt as a basis: and, secondly, what system of numerical multiplication and aliquot sub-division of such unit for measures of length, and of its derivative units of area, of capacity, and of weight (for these all refer themselves naturally and easily to the unit of linear measure, or at least ought to do so) is most advantageous—either in a great mercantile community like our own, or for the great mass of mankind in the ordinary transactions of life. And it cannot be too strongly impressed, and too perseveringly borne in mind, that these two questions stand in no natural and necessary relation to each other, but are perfectly independent. We may resolve, with perfect logical consistency, either to toss aside our present system *in toto*, and adopt the metrical one in preference; or to retain our fundamental unit (the Imperial foot or yard), and decimalize our system of denominations; or, lastly, by a slight, and, practically speaking, imperceptible change in our present standard, to bring it into conformity with our views of theoretical perfection (which, I shall show, may be done). We may, too, retaining, all the convenience of our existing denominations (*so far as they are convenient*) superadd to them, by permissive legislation, the additional convenience of a decimal system for facility of calculation: relying on its holding its ground if really affording such facility, or working its way into general use, and ultimately driving out the old system, if found by the mass of the population to be practicably preferable. This last is the course I would myself prefer, and I think it best to say so in the outset, lest those who may take a contrary view should imagine a foregone conclusion to be urged upon them under the semblance of free inquiry.

(3.) It is unnecessary, of course, to observe that, the measurement of length being required for almost every purpose of construction as well as for every intelligible statement of the sizes of material objects, the lengths of journeys, the distances of places, &c.—renders indispensable the recognition, in every community, of some common standard, some well-known and identifiable *unit*, by whose repetition great, and by whose aliquot subdivision small lengths, distances, sizes, &c., may be expressed in words and numbers. The common sense of mankind; moreover, would naturally point, in the selection of such unit, to some object of common occurrence, of moderate linear dimension, and of which individual exemplars differed but little, or, if possible, not at all in this respect; so that appeal might at once be made to such exemplar in case of a question arising as to the length of any object stated to contain a given number of such units or its aliquots. A very moderate experience would however suffice to convince anybody that among natural objects of the same kind, even those most common, perfect identity of length, of breadth, of thickness, any more than of weight, is never observed—even a close approach to it rarely—and a very close one extremely so. Still, with all drawbacks so arising on the adoption of a natural standard, the first rude demand for such a standard would be easily enough satisfied, and that in two ways, viz.: 1st, by actually fixing upon some individual among all the existing objects of the sort selected, to the exclusion of others—or, 2dly, by the very natural, though somewhat more refined conception of an ideal medium, or *mean* among a very great multitude of such objects, such as might be regarded as neither unusually great nor unusually little ones of their kind.

(4.) Among objects of common occurrence, the human person, or some distinct member of it would be most likely to claim the attention of mankind as affording a standard of measure; if only for the very obvious reason that the relation of the sizes of material objects to that of man mainly determines his facility of handling, or otherwise applying them to human uses. Accordingly, the height of a full grown person, the length of his arm, his fore-arm (ulna or ell), his foot, his hand, his ordinary step, &c., would present, and is well known to have presented itself among almost all communities of mankind to their choice for this purpose.

And so, among all nations whose measures have been handed down to us, we find in speaking of the unit of length, some members of the human person designated. Thus, the bed of the gigantic king of Basan is related to have measured eight cubits in length "after the cubit (*i.e.*, the fore-arm) of a man." The height of Goliath the Philistine was "six cubits *and a span*." The bow of Pandarus, described by Homer, was formed of the horns of an Ibex, which grew out sixteen *palms* (or hand-breadths) from his head. The Romans reckoned their distances by intervals of 1000 *paces* (*millia passuum*) whence our *name* for a mile, though differing widely in reality. If, however, we may judge from the great diversity in the actual lengths adopted under the common name of "a foot" as the standards of different nations, we shall see reason to believe that the typical foot selected was usually that of an individual—some Chief, King, or High Priest, who could claim pre-eminence among them as a man *par excellence*, and who would seem to have been generally above the average stature. Thus we find the Roman foot equivalent to 11·6 of our inches; the English to 12; the Greek to 12·1; the French to 12·8; and the Egyptian or "Drusian" to 13·1—all of them (especially the two last) in excess of the real length of the foot of a well-proportioned man of medium stature (say 5ft. 10in.) which does not exceed 10¾, or at the most 11 inches.

(5.) Another class of objects, which, from the universality of their occurrence in vast numbers, and their general uniformity of dimension, would naturally occur as unit types, available for the measurement of small lengths, or for the small aliquots of a larger unit, has been found in the cereal grains of most common use, and of these, the barley corn, and the rice grain, have found the preference. Our inch, for instance, has been defined in an old statute (now repealed) as the length of three grains of barley, taken from the middle of the ear, and placed end to end. And in a somewhat similar manner have been derived from those cereals the smaller sub-divisions of the Hebrews and Hindoos; while the larger have, in these, as in other nations, originated in parts of the human person.

(6.) It is very evident, however, that types of this kind admit of no precise and rigorous identification or intercomparison. The medium stature of a man is very different in different countries. That of an adult French conscript for instance, is (or at least was in 1817) 5ft. 4in., as concluded from the measurement of 100,000 individuals, while the Belgian type, or mean adult stature, has been placed at 5ft. 7in. ·8, and that of a Lancashire non-manufacturing labourer, as high as 5ft. 10¾in. So great a discordance as a result of local and secondary circumstances, is of course fatal to the pretensions of the human person as a natural type. So again of the cereals. The difference of soil, climate, and cultivation must produce, and does in fact produce very great variety in the medium size of grain grown in different countries, and in different years: so that, even supposing them to be measured by millions, the mean results would be found to differ too much for the object in view. And the same kind of objection holds good against having recourse to any kind of medium magnitude, among multitudes of objects of a like species which occur in nature. Such must, of necessity, be chosen among organic structures of the animal or vegetable kingdom (for among inorganic masses of whatever kind, nature presents no instance of a *mean* or typical magnitude, as distinct from the *average* of a number accidentally assembled, which may differ to any extent from an average similarly taken of an equal number elsewhere collected). And among the former classes of objects, even were it possible to assemble and measure them in sufficient numbers to afford a true *typical mean*, we should have no security for its identity in different ages and climates.

(7.) We are driven then, in our choice of a *universal* standard to the selection, either of some individual object, (if such there be) natural or artificial, imperishable in its nature, unsusceptible of variation by lapse of time or decay, and indestructible by accident—or else, to some ideal or resultant length or magnitude (if such there be), susceptible by its definition of being as it were translated into a material expression, and marked out as the result of some process which we are sure will, in all ages and places reproduce the same identical result. And besides these qualities of invariability, indestructibility and identical reproducibility it ought to possess some obvious claim to *general* acceptation as of common interest to all mankind, or at least to all the civilized portion of it: an interest from which national partialities and rivalries should be altogether excluded.

(8.) The individual human type is at once excluded

by these conditions. Supposing the foot of the most remarkable person who ever lived to be marked out on steel or adamant, it would be at the mercy of fire, earthquake, loss in political convulsions, and a hundred other forms of destruction or disappearance, without the possibility of reappeal to the original form. Of human works, the most permanent, no doubt, and the most imposing as well as generally interesting and respected, are those mighty monumental structures which have been erected as if for the purpose of defying the powers of elementary change. Take the vastest of them—that to which appeal has been often made for this very purpose—the great pyramid of Cheops. When built it was 481 ft. in height, and the square area of its base was 764 ft. in the side. The height is now only 451 ft. and the side of the base only 746; and the sole means by which we are now enabled to determine the original height consists in a block of the exterior marble casing which will in all probability disappear in the hands of "the curious" within the next century. Nature presents to us but one material *object* which combines all the requisites enumerated, and combines them all in perfection—viz.: the globe itself that we inhabit. And in that globe we find only two naturally-defined lengths which unite the requisites of individuality to identify them under every change of human relations and even of geological revolutions and catastrophes, and of universality, so as to stand in the same relation to both hemispheres and to all meridians—viz.: the earth's polar axis, and its equatorial circumference. For the latter, the equatorial *diameter* might be more advantageously substituted: but that we have good reason to believe the equator to be not strictly circular, but in some degree elliptic, the proportion of its greatest and least diameters not being yet precisely known, though very much nearer to equality than that of the equatorial and polar diameters. This however would not prevent its *mean* equatorial diameter from being assumed in preference to its circumference, were not the polar axis, for very obvious reasons, preferable to both. Of the latter, and indeed of all three (thanks to the elaborate geodesical surveys which have been made within the century last elapsed), we possess a knowledge so precise as to render them perfectly available for our purpose.

(9.) Of lengths which exist not marked by the dimensions of any material object, but which are defined by the nature of things and by physical relations, and which are susceptible of exact determination and of being marked off on a scale, and so of becoming materialized for practical reference; there have been proposed only three which can be considered theoretically, and of these only one practically available. These are, 1st, the velocity of light or the space travelled over by light in some definite time (say the ten-millionth part of a second, which would give a modulus of about 100 feet); 2dly, the length of an undulation of a ray of light of some definite refrangibility—a length so minute as to require multiplication a million-fold to give a modular unit; and 3dly, the length of a pendulum vibrating seconds under certain definite and normal circumstances—or rather that of an ideal seconds-pendulm supposed to be placed at the extremity of the earth's polar axis. To this is in effect equivalent, and derivable from it, as a mere arithmetical conclusion, the space fallen through by a heavy body on the same place by the earth's attraction in a second of time. The modulus so obtained is therefore a measure of the earth's total attractive power (independent of centrifugal force arising from its rotation), as that derived from the length of its diameter is of its total bulk, and equally unalterable and universal. As for the other two which depend on the nature of light, the difficulty and delicacy of the processes they would involve render all idea of resorting to either of them purely visionary.

(10.) The linear dimensions of the earth then, on the one hand, and the linear measure of its attractive force embodied in the pendulum on the other, are the two, and, so far as we can see, the only two available sources of the invariable and universal standard length which we seek. And it is curious to observe that while the French after considering both of them threw aside the pendulum in favour of the metre (or ten millionth of the meridian quadrant); the English on the other hand, by the Act of Parliament in 1824, which repealed the old statute already alluded to (and so threw aside the principle of resorting to an organic type) did in effect, at that time, adopt the pendulum as their ultimate resort. For while that act declares that a certain metallic bar made by Bird in 1760 when at the temperature of 62 Fahr. should, without any further reference to its origin, be considered the standard yard of the British empire, it provided for its recovery and reproduction in case of the total destruction or loss of it and all its authentic copies and facsimiles, by a declaration that its length is 36 inches, *such* that 39·13929 of them are equal to the length of a pendulum vibrating seconds *in vacuo* and at the sea-level, in the latitude of London. The report of the French commissioners also in 1798 which led to the enactment of the metrical system, is careful to state that in the event of the total loss or destruction of all material representatives of the metre its value would be easily recoverable from a numerically specified relation between its length and that of the pendulum vibrating seconds at Paris, which had been determined with great accuracy by Borda, one of the commissioners. So that, practically speaking, in the event of the total destruction, by political convulsions, of every authentic yard and metre (supposing any written record of our existing knowledge to survive them) the metre would have been recovered, not by the laborious and costly process of remeasuring the French meridian arc, but by the infinitely more summary one of a precise repetition of Borda's experiments and the exact re-application of all his corrections and reductions.

(11.) For the reproduction of the English yard, a similar repetition of those experiments *in* London which led to the adoption of the number 39·13929 in. as the measure of the pendulum would, in such an event, no doubt have been, at that epoch, resorted to; though in departure from the wording of the act, which speaks of a pendulum vibrating seconds, not *at* but *in the latitude of* London: a very different thing, as General Sabine has pointed out in his "*Account of Experiments to determine the figure of the Earth by means of a Pendulum vibrating Seconds in different latitudes.*" For the object would have been then, as it really was on the occasion of the actual destruction of the parliamentary standard in 1834, not to *produce* a theoretically *better*, but as far as possible to *reproduce* the same *identical* length by the most summary process; without undertaking circumnavigatory voyages, or entering on any theoretical discussion. The new act necessary for legalizing the standard so arising would probably have sanctioned this procedure, and we should have thenceforward had a standard of a purely local character, assuming for the fundamental basis the individual seconds pendulum *in* London.

(12.) This, however, is not now the case. On the destruction of the standard of 1760 by the burning of the Houses of Parliament, the new standard was constructed, not by any measurement of the length of the pendulum (for in the ten years elapsed since 1826 very grave doubts had been raised, or rather very serious sources of error pointed out in the processes used for the purpose on the former occasion)—but, by an assemblage and most careful comparison of all the scales and standards of any authority which could be got together—resulting in the production of one primary and a great many secondary standards, in all human probability absolutely identical with that destroyed. The act, moreover, (of 1855) which constituted that one our legal yard, and named the others in a certain order as its successors in the event of its destruction or loss, omitted the clause identifying its length with any numerical multiple of the pendulum. In fact, then, our yard is a purely individual material object, multiplied and perpetuated by careful copying; and from which all reference to a natural origin is studiously excluded, as much as if it had dropped from the clouds. Apart, then, from the extraordinary pains taken in its construction, and from the singularly fortunate but at the same time purely accidental coincidence which I shall presently mention, it has no pretensions whatever to be regarded as a scientific unit.

(13.) Let us now consider the claim which the pendulum, in the abstract, as a measure of the earth's gravitation, can advance for its reception as a fundamental and universal standard of length (and here, incidentally it may be remarked that, *as a length*, it is not more inconvenient than the metre, being within about a quarter of an inch the same).* One of the reasons assigned by the French *Savans* for their rejection of it in favour of the metre, and, as would appear, the only one which weighed with them (for their other reason ostensibly advanced is a mere appeal to the political passions of the time) was the dependence of the length of the pendulum

* The metre has this inconvenience, as compared with the yard —that while the latter can be readily extemporized by a man ot ordinary stature (and often is so in practice) by holding the end of a string or ribband between the finger and thumb of one hand at the full length of the arm extended horizontally sideways, and marking the point which can be brought to touch the centre of the lips (facing full in front); the former is considerably too long to afford the same facility.

on the time of its vibration; as if the 86,400th part of a day which we call a second of time were not as definite and as invariable a quantity as the 100,000th part which, in their rage for decimalization, they proposed to call one; and as if they might not have fixed on a pendulum vibrating 100,000 times in a day (which would have given a very near approach to our yard). But their stumbling-block was the introduction of an extraneous element, *time*, at all, into the subject: as if the length of the day were not as much an invariable, universal, and physical element as the dimensions of the earth or its gravitation. But in this they seem to have overlooked the fact that their adoption of the quadrant of a meridian for the base of their system does really admit this extraneous element, time, into that system, though in a much more insidious way. For the total bulk or mean radius and the total mass or gravitating energy of the earth remaining the same, the ellipticity of its meridians, and therefore their absolute length, depends on the period of its rotation or the length of the day. The same objection, to be sure, if it be one, would equally apply to the adoption of the polar axis, or the equatorial diameter of the earth; and the only way to exclude all ideas of time and force from a metrical system, and render it *purely* metrical, *i.e.*, dependent on geometrical magnitude alone, would be to take for a fundamental unit the radius, diameter, or circumference of a sphere, or the side of a cube, equal in volume to that of the earth. And perhaps were a *tabula rasa* made; were the ground totally unoccupied and the whole matter to do over again, this would be as good a unit as could be proposed.

(14.) But the true objection to the choice of the pendulum for a universal unit of measure lies, not in any metaphysical and abstract considerations of this kind; but in the uncertainty which prevails, and must necessarily always prevail as to the true length of that normal or ideal pendulum which shall stand equally related to the whole globe, and from which the mean length corresponding to any assigned latitude can be calculated: that is to say, the length of a pendulum which would swing seconds at the pole of the terrestrial spheroid— an uncertainty which, as I shall proceed to show, must affect the result of every attempt to deduce it with the precision the subject requires from experiments made on the surface of our planet: however refined the methods employed, and however numerous and diversified the geographical stations at which they may be instituted.

(15.) In practice, the mean length of the polar or equatorial pendulum is concluded from an assemblage ot the observations of the times of oscillation of one and the same invariable pendulum at a multitude of geographical stations in all accessible latitudes in both hemispheres: no two combinations agreeing in giving the same precise length, by reason of the local deviations of the intensity of gravity due to the nature of the soil, and the configuration of the ground immediatety beneath and around the places of observation. Now, since the pendulum cannot be observed at sea, the whole sea-covered surface of the globe is of necessity excluded from furnishing its quota of observations to the final or mean conclusion. And the influence of this, it should be observed, is not self-compensating as that of local inequalities of mere density on land would be, but tells all in one direction. For water being, on the average, not more than one-third the weight of an equal bulk of land (such land as the earth's surface consists of) and only $\frac{2}{11}$ of the mean density of the globe, the force of gravity at the surface of the sea is less than at the sea-level on land by the attractive force of as much material taken at twice the specific gravity of water, or at $\frac{4}{11}$ths that of the globe, as would be required to raise the bottom to the surface. Supposing then the difficulty of observing the pendulum at sea overcome, and that the whole surface of the globe were dotted over with stations of observation equally distributed over sea and land, from whose intercomparison it were required to derive the mean co-efficient of terrestrial gravitation, or the mean length of the polar pendulum; it is evident that the sea stations would everywhere conspire to give a less result than the land. According to Dr Young (Phil. Trans., vol. cix., page 93) the attraction of an extensive flat mass of any thickness on a point in the middle of its surface is three times that of a sphere of the same materials having that thickness for its diameter. And from this it is very easy to conclude that, supposing the sea to have a mean depth of four miles (which seems not improbable) the mean defalcation of gravity at its surface, due to the deficiency of attracting matter, would be three times the attraction of a sphere four miles in diameter and $\frac{4}{11}$ths of the earth's mean density—that is by a simple calculation $\frac{1}{1833}$, or rather less than one 1800th part of the whole attraction of the earth—a fraction far too large, as well as far too uncertain in its amount either at any given spot or in general, not to vitiate irremediably any conclusion as to the ultimate result of the operation.

(16.) Similarly, if we look to the reductions to the sea level necessary for stations in the interior of continents, we shall find that they depend, partly on the diminution of gravity due to the *height above* the sea-level, or to the increase of distance from the earth's centre, which always tells in *diminution* of gravity; and partly on the protuberant matter, be it mountain or elevated table-land immediately beneath and around the pendulum, which always tells in favour of *increased* gravitation. The former portion is rigorously calculable, and therefore need not trouble us, but the latter is in an extreme degree uncertain in particular localities, and in a general estimate falls very short of compensating for the sea-deficiency. For the mean height of the European continent is only 1342 feet; of Asia 2274; of North America 1496; and of South America, 2302. The mean is 1840 feet, or rather more than a third of a mile, which, on the same principle of reckoning, would be equivalent to about $\frac{1}{15000}$th part only of the total gravity, which has to be reduced to one-third of its amount, or to 1-45000th, inasmuch as the proportion of land to water over the whole globe is only that of 51 to 146, or about 1 to 3. This is the mean effect of the elevated matter to *increase* gravitation. That of mere *elevation* above the sea-level to the height of $\frac{1}{3}$ of a mile (similarly reduced) is, however, one 36000th in the opposite direction, or to diminish it—and the difference or one 180,000 of the whole is effective *not to compensate but to add to* the sea-deficiency.

(17.) To obtain the real length of the normal pendulum then we must go out of our own globe, and ascertain the true co-efficient of gravity from astronomical facts; and, as the only one available for the purpose, compute the distance fallen through by the moon in a second of time towards the earth from a tangent to her orbit. This, it is evident, is independent of the influence of those local inequalities which affect the pendulum measurements. But, on the other hand, it must be remembered—1st, That our knowledge of the distance in question depends on our previous knowledge of the moon's distance, which, in its turn, depends on that of the earth's diameter, and therefore presupposes the metre to be *accurately* known. For any *aliquot* error in the metre will produce an equal *aliquot* error in the moon's distance estimated in metres, and therefore also in the linear deflection per second from the tangent to the orbit. 2d, That this linear deflection, or approach of the moon to the earth in one second of time, is the result of the joint attraction of the earth on the moon and of the moon on the earth, and is in effect the sum of the spaces fallen through by the moon towards their common centre of gravity, in virtue of the earth's attraction, and by the earth towards that point in virtue of the moon's. Now the mass of the moon is about one 88th part of that of the earth, so that one 88th part of the force that draws them together is due to the moon. By so much then must the space fallen through be diminished, to get that due to the earth's alone. Suppose, now, that the moon's mass assumed should be in error by a 50th part of its whole amount—(and Laplace's estimate of it differs by as much from that at present received)—and we shall find ourselves landed, from this cause of uncertainty alone, in an error to the extent of nearly one 4000th of the quantity sought.

(18.) Lastly, our knowledge of the moon's mass is mainly derived from its effect in producing the phænomenon of nutation, which it does through the medium of the earth's ellipticity, so that not only the dimensions, but the figure of the earth are thus mixed up in our attempt to derive the length of the normal pendulum from the moon's motion.

(19.) I cannot but consider then that the uncertainty of the one mode of obtaining the length of the normal pendulum, and the non-independence of the other, unfit it for being received as the ultimate scientific basis of a universal standard; whatever merit it may possess in an abstract and metaphysical point of view—and that the true and only practical use of the pendulum in relation to such a standard is the ready, cheap, and perfectly unobjectionable means its measurement, at a determinate spot and under defined circumstances, affords of recovering it when lost, by the recorded statement of its length in terms of such standard.

(20.) The causes of uncertainty which tell with such very appretiable effect on the local determination of the force of gravity by the pendulum, have little or no influence on the local curvature of the surface of equilibrium, and absolutely none on the measures of large arcs of the meridian. Suppose, for example, a sea of four miles in depth, and of great extent, to cover one part of the earth's surface. Its surface water will gravitate less by one 1800th part of its proper weight, owing to the deficiency of attracting matter below it; and, the diminution of gravity growing less and less in descending (being proportional to the height of a particle above the bottom), the whole weight of the column of water vertically above a given spot will be diminished by one 3600th part, so that to maintain the equilibrium, one 3600th part of four miles, or one 900th of a mile, *i.e.*, about six feet of additional water, must be heaped on: a mere infinitesimal of the radius of curvature of its surface, which is that of the earth itself.

(21.) Let us now see how far the French metre, as it stands, fulfils the requirements of scientific and ideal perfection. It professes to be the 10,000,000th part of the quadrant of the meridian passing through France from Dunkirk to Formentera, and is therefore, scientifically speaking, a local and national, and not a universal measure. The earth's equator is not a perfect circle, but slightly elliptic, and the meridians of places differing in longitude are therefore not all of the same length. The difference, however, is so trifling (the ellipticity of its equator being not more than a thirtieth part of that of its meridian) that to raise an objection against the practical reception of the metre, either *per se*, or as a substitute for the yard, on this score, would savour of hypercriticism. A more serious objection is the choice made of the circumference of the meridional or generating ellipse of the terrestrial spheroid in preference to its axis of revolution. This is a blemish on the very face of the system—a sin against geometrical simplicity. Still, were the length of the metre as determined by the French geometers rigorously exact, or correct within limits which the much more extensive measurements of meridian arcs since made elsewhere than in France have proved to be attainable, this would be only a matter of regret, and could hardly, of itself, be drawn into an argument for its rejection. But this is far from being really the case. The metre, as represented by the material standard adopted as its representative, is too short by a sensible and measurable quantity, though one which certainly might be easily corrected. To show this it will be necessary to enter into some detail.

(22.) In effect, that standard is declared, in the Annuary of the Bureau des Longitudes, to be equal to 39·37079 British imperial standard inches. The quadrant of the French meridian then ought, if this be correct, to be 393,707,900 such inches, or 32,808,992 feet. And by whatever aliquot part of its whole length the true quadrant exceeds this, by that same aliquot of *its* length is the metre, so stated, erroneous.

(23.) Mr Airy, by a combination of the whole series of meridian arcs whose measures had been obtained in every part of the globe in 1830, was led to conclude for the value of the minor or polar axis of the terrestrial spheroid, 41,707,620 feet; while the late Professor Bessel, pursuing a course similar in its general principle—that is to say, using all the measured arcs, great and small, in combination one with another, and taking the most probable mean among the (necessarily) discordant results, obtained by combining them two and two—arrived at a value very slightly different, viz., 41,707,314 feet. The mean of these gives, as the result of this mode of procedure, 41,707,467.

(24.) Quite recently, M. Schubert in a very elaborate memoir which appears as part of the 1st vol., 7th series, of the Memoirs of the Petersburg Academy, has pointed out the inconvenience, and necessarily discordant results which the combination by pairs of a multitude of small arcs, each of itself insufficient to afford any precise measure of the ellipticity, affords; and assigned his reasons for restricting the inquiry in the first instance into the length of the polar axis, as an element unique in itself, and common to all the meridians: deducing it separately from each of the most extensive arcs, the Russian, the Indian, and the French, each taken independently;—comparing the three values so obtained, and thence concluding the final result. In this manner he obtains the following three values of the axis, viz.:—

From the Russian arc	(of 25° 20′	in extent)	41,711,019·2 feet.
,,	Indian ,,	(of 21° 21′	,,)	41,712,534·2 feet.
,,	French ,,	(of 12° 22′	,,)	41,697,496·4 feet.

In concluding for these a mean, or final value, M.

Schubert however, arbitrarily, and as I think quite indefensibly, rejects altogether the result of the French arc, and assigns to the Russian double the weight of the Indian; a mode of precedure in which he will find, I presume, few to agree with him. A much fairer, indeed the only fair way to treat them, is obviously to ascribe to each of the separate results in taking the mean, a weight proportional to the total extent of the arc, and this gives for the length of the axis 41,708,710·0 feet. Comparing then the final results of the two modes of procedure we find,

From the former, ... 41,707,467 feet.
And from the latter, ... 41,708,710 „

which differ only by 1243 feet, or less than $\frac{1}{4}$ of a mile—so that their mean or 41,708,088·5 f. is in all probability within a furlong, or one part in 64,000 of the truth.

(25.) From each of the great arcs of Russia and India, M. Schubert then obtains a separate value of the equatorial or the larger axis of the elliptic meridian to which it belongs; and by a similar treatment of the arc of Peru, which, lying under the equator, is especially favourable for the purpose, he obtains a third value of the equatorial diameter. The three diameters of the equatorial ellipse thus obtained, with the angles they include at the centre (which are the differences of longitude of the respective meridians, and which are as favourably arranged for the purpose as the nature of the case seems to admit), suffice for the determination of the major and minor axis of the equator, regarded as an ellipse, and the longitudes in which they lie, viz.:—

Axis major = 41,854,800 feet, in long. 38° 44′ E. from Paris (one end falling about half-way between Mount Kenia and the east coast of Africa, the other in the middle of the Pacific Ocean).

Axis minor = 41,850,007 feet, in long. 128° 44′ E. from Paris (one end falling on Waygiou, one of the Molucca Islands, and the other at the mouth of the Amazon River), giving an ellipticity of one 8880th, or about one-thirtieth part of that of the meridians as already stated.

(26.) The figure of the equator, and its dimensions thus obtained, the exact equatorial diameter corresponding to any given longitude is easily calculated. And by comparing this with the polar axis, the precise ellipticity of the meridian for that longitude may be computed. And executing this computation for Paris, M. Schubert finds $\frac{1}{296}$ for the ellipticity of the French meridian.

(27.) With these data, viz., a Polar axis of 41,708,088 feet, and an ellipticity of $\frac{1}{2.96}$ which certainly may lay claim to greater precision than anything previously obtained, I shall now proceed to calculate the true length of the quadrant of the French meridian, for which purpose the following very simple and convenient formula may be used,* viz.:—

$$Q = \frac{\pi}{4} A (1 + 2m + 9m^2 + 38m^3)$$

* For the present purpose it is necessary to carry out the calculation to the cube of the ellipticity—but in cases where the *square* of that fraction may be neglected, the following simple rule for finding the circumference of an ellipse is worth remembering. On the longer axis of the ellipse describe a circle, and between this and the ellipse, describe a small circle having its centre in the prolongation of the minor axis, and touching the ellipse externally, and

in which Q represents the length of the quadrant required, A that of the polar axis, π the circumference of a circle whose diameter is 1, and *m*, *one fourth part* of the fraction expressing the ellipticity, or in this case $\frac{1}{1184}$.

Executing the calculation the result is...32,813,000 feet.
Substract 10,000,000 metres = 32,808,992

Remain, excess.............. 4,008

for the excess of the true quadrant over that assumed as the basis of the metrical system, that is to say, one 8194 aliquot part of the whole, or one 208th of an inch on the whole metre, which is therefore the quantity by which the French standard is actually too short.

(28.) It must not be denied that this is a very wonderful approximation, and in the highest degree creditable to the science, skill, and devotion of the French astronomers and geometricians who carried on their operations under every difficulty, and at the hazard of their lives in the midst of the greatest political convulsion of modern times. And adopted as it is over a large portion of Europe; were the question an open one what standard a new nation, unprovided with one, unfettered by usages of any sort, and in the absence of any knowledge of the existence of the British yard, should select; there could be no hesitation as to its adoption (with that very slight correction above pointed out—which would in no

the circumscribed circle internally. The circumference of this small circle is the difference between those of the ellipse and of the larger or circumscribing circle.

way interfere with its practical use—a correction which the French themselves might, under such circumstances, consent to adopt). But the question now arising is quite another thing, viz.: whether we are to throw overboard an existing, established, and, so to speak, ingrained system—adopt the metre as it stands, for our standard—adopt moreover its decimal sub-divisions, and carry out the change into all its train of consequences; to the rejection of our entire system of weights, measures, and coins. If we adopt the metre we cannot stop short of this. It would be a standing reproach and anomaly—a change for changing's sake. The change, if we make it, must be complete and thorough. And this in the face of the fact that England is beyond all question the nation whose commercial relations, both internal and external, are the greatest in the world, and that the British system of measures is received and used, not only throughout the whole British empire (for the Indian "Hath" or revenue standard is defined by law to be 18 British imperial inches) but throughout the whole North American continent, and (so far as the measure of length is concerned) also throughout the Russian empire; the standard unit of which, the Sagene, is declared by an imperial ukase to contain *exactly* seven British imperial feet, and the Archine and Vershock precise fractions of the Sagene. Taking commerce, population, and area of soil then into account, there would seem to be far better reason for our continental neighbours to conform to *our* linear unit could it advance the same, or a better *à priori* claim, than for the move to come from our side. (I say nothing at present of decimalization).

(29.) Let us see then how this part of the matter stands. Taking the polar axis of the earth as the best unit of dimension which the terrestrial spheroid affords (a better *à priori* unit* than that of the metrical system) we have seen that it consists of 41,708,088 imperial feet—which, reduced to inches, is 500,497,056 imperial inches. Now this differs only by 2944 inches, or by 82 yards from 500,500,000 (five hundred million and five hundred thousand) such inches—and this would be the whole error on a length of 8000 miles which would arise from the adoption of this precise round number of inches for its length, or from making the inch, so defined, our fundamental unit of length. Suppose, then, that any length were proposed in English measure, and we desire to know what decimal fraction such length were of the earth's axis. We have only to express it in inches and decimals, and from the number so stated take off its thousandth part (a calculation involving only the writing down the number twice over, removing the figures of the

* A writer in *Quesneville's Moniteur Scientifique*, No. 163, v. 736, argues that *itinerary* measures ought to be based on the *circumference* of the globe and not on its *axis*—by reason that the decimal principle of sub-division, if carried out, would apply to the decimal graduation of the quadrant—*adding* that "the greatest advantage of the French system is in reality its decimal division"—but *forgetting* to add that the decimal division of the quadrant *was* introduced in France, but *was abandoned by common consent even in France*, and can never be reintroduced. In the "*Mondes*" (Suppl. 38, p. 616) the same argument is advanced, and the same answer applies.

under line three places to the right and subtracting), and the thing is done, and *vice versâ.** Suppose now the same length stated in French metres, and we would ascertain what decimal fraction it is of a quadrant of the French meridian. The number of metres assigned must be divided by 8194 either by a long division sum or by the use of a table, before the proper number to be subtracted can be found. Which then is the shorter process? and which, both scientifically and practically, the preferable unit?

(30.) If we are to legislate at all on the subject then, the enactment ought to be to increase our present standard yard (and of course all its multiples and submultiples) by one precise thousandth part of their present lengths, and we should then be in possession of a system of linear measure the purest and most ideally perfect imaginable. The change, so far as relates to any practical transaction, commercial, engineering, or architectural, would be absolutely unfelt, as there is no contract for work even on the largest scale, and no question of ordinary mercantile profit or loss, in which one *per mille* in measure or in coin would create the smallest difficulty. Neither could it be doubted that our example would be

* Strictly speaking for the conversion and reconversion we should *subtract* one 999th and *add* one 1000th. But the difference is only one part in a million which can never be of the slightest importance. *Per contra* the conversion of the metre according to the process here stated leads to a result which, though exact in parts of the *French* meridian, is erroneous in parts of the *mean terrestrial* meridian by a considerably larger proportional part, and *this* is what we really want to know.

very speedily followed both in America and Russia, so soon as the reason of the thing and the trifling amount of the change came to be understood. And even without legislation the relation between the proposed new or *geometrical* measure and the imperial ones is so simple and striking—fixing itself so easily in the memory, and the conversion from one to the other so ready, that, *were there no other reason*, it might almost be questioned whether it would be worth while to make the change.

(31.) But there *is* another reason, and I think a decisive one. Hitherto I have said nothing about our weights and measures of capacity. Now, as they stand at present nothing can be more clumsy and awkward than the numerical connexion between these and our unit of length. A grain is defined as the weight of distilled water, so that 252·724 of such grains at the freezing temperature, or 252·46 at that of 62° Fahr. which is the standard temperature of our imperial yard, shall fill a cubic inch. Of such grains, so defined, the pound contains 7000, the ounce 437½, and the gallon of water at 62°, 70,000. According to this system, the cubic foot of water at our standard temperature weighs 997·145 oz., falling short of 1000 oz. by very nearly 3 oz. However tempting this approximation might appear, still, in the absence of any more cogent reason, the commissioners who recommended our system of weights and measures legalized in 1824 forbore to recommend such a change in the ounce (about 1⅓ grain) as would have brought it about; though the rule that a cubic foot of water weighs 1000 ounces is still handed down as a rough and ready way of converting cubic measure into weight. But were we to adopt the *geometrical* instead of the present imperial standard—the linear foot being increased by one thousandth, the cubic foot would be increased by three times that aliquot, or would become 1·003 times our present cubic foot—and so would make up just the deficient three ounces, or at least so very nearly that a legislative change in the ounce, increasing it by only one part in 8000, or by one 18th part of a grain, would bring everything into decimal coincidence, by making the ounce and the cubic foot the links of connexion between weights and measures instead of the grain and the cubic inch, as at present. As regards our measures of *capacity*, the connexion would be equally consecutive, as a decimal one, between the cubic foot and the half pint, which for the purpose in view, ought to have a distinct name (such as a "*tumbler*," or a "*rummer*," or a "*beaker*")—and which would contain exactly one 100th part of a cubic foot,—with whatever liquid or solid matter it might be filled. And thus the change which would place our system of linear measure on a perfectly faultless basis, would at the same time rescue our weights and measures of capacity from their present utter confusion, and secure that other advantage, second only in importance to the former, of connecting them decimally with that system on a regular, intelligible and easily-remembered principle; and *that* by an alteration practically imperceptible in both cases, and interfering with no one of our usages or denominations.

(32.) On the subject of decimalization, it will be gathered from what I have said that I would make any decimalized denominations which anybody might agree to buy, sell, or contract by, permissive. There seems to be a doubt whether such is now the case, and if so the law should I think be altered. But I would leave untouched all our present *denominations* and their relations to the standard—and the only new measure I would legalize would be a "module" (or some other name *at present unoccupied*) of 50 geometrical inches being the ten millionth of the polar axis, or its half, the "geometrical cubit" of 25 such inches—leaving its use quite voluntary.

COLLINGWOOD, *Sept.* 30, 1863.

ADDENDUM.

(33.) Since the foregoing remarks were written my attention has been called by the Astronomer Royal to a very elaborate memoir by Captain Clarke, in vol. xxix. of the Memoirs of the Royal Astronomical Society, whose conclusions, though differing from those of M. Schubert in some particulars (as in making the equator more elliptic) yet, so far as the present subject is concerned, tend in the same direction, and that, as regards the aliquot error of the metre, even more strongly.

(34.) Captain Clarke assigns for the three axes of the earth the following values:—

Polar axis	41,707,536 feet.
Or in inches	500,490,432.
Longer equatorial axis	41,852,970 feet.
Shorter do. do.	41,842,354 "

Longitude of the vertex of the longer axis = 13° 58′ 30′ east—or 11° 35′ 15″ E. of Paris) whence it is easy to conclude as follows:—

Diameter of equator in the longitude of Paris...41,852,695 feet.
Ellipticity of the Paris meridian.......................... $\frac{1}{288 \cdot 2}$ say $\frac{1}{288}$

(35.) Calculating now the quadrant from this ellipticity, and from Captain Clarke's polar axis, we find it 32,814,116 feet, which exceeds ten million metres by 5124 feet, being in excess of that above found (4008) by 1116 feet; and corresponding to an aliquot error of one part in 6404, or on the metre itself to one 163d part of an inch. The aliquot error in our "geometrical yard" is also somewhat increased by the adoption of this polar axis, viz., to one part in 52,310, or to about one 1453d part of an inch on the yard.

(36.) As this memoir of Captain Clarke contains by far the most complete and comprehensive discussion which the subject of the earth's figure has yet received, and must be held as the ultimatum of what scientific calculation is as yet enabled to exhibit as to its true dimensions and form—this conclusion will of course be considered to supersede that arrived at in the foregoing pages.

COLLINGWOOD, *Oct.* 11, 1863.

P.S.—Some slight subsequent corrections made by Capt. Clarke in his calculations, founded on data quite recently published, make the polar axis approximate *still more nearly* to 500,500,000 inches.

XI.

ON ATOMS.*

"I sing of atoms."—*Rejected Addresses.*

DIALOGUE.—Hermogenes et Hermione interloquuntur.

HERMIONE.—What strange people those Greeks were? I was reading this morning about Democritus, "who first taught the doctrine of atoms and a vacuum." I suppose he must have meant that there is such a thing as utterly empty space, and that here and there, scattered through it, are things called atoms, like dust in the air. But then I thought, "What *are* these atoms?" for if this be true, then, these are all the world, and the rest is—nothing!

Hermogenes.—Yes. That is the natural conclusion: unless there be something that does not need space to exist in; or unless there be *things* that are not mate-

* From the *Fortnightly Review*.

rial substances; or unless space itself be *a thing:* all which is deep metaphysic, such as I am just now rather inclined to eschew. But, dear Hermione, how am I to answer such a host of questions as you seem to have raised—all in a breath? The Greeks! Yes; they were a strange people—so ingenious, so excursive, yet so self-fettered; so vague in their notions of things, yet so rigidly definite in their forms of expressing them. Extremes met in them. In their philosophy they grovelled in the dust of words and phrases, till, suddenly, out of their utter confusion, a bound launched them into a new sphere. There is a creature, a very humble and a very troublesome one, which reminds me of the Greek mind. You might know it for a good while as only a fidgety, restless, and rather aggressive companion, when, behold, hop! and it is away far off, having realized at one spring a new arena and a new experience.

Hermione.—Don't! But a truce to the Greek mind with its narrow pedantry and its boundless excursiveness. The excursiveness was innate, the pedantry superinduced —the result of their perpetual rhetorical conflicts and literary competitions. I have read the fifth book of Euclid and something of Aristotle; so you need not talk to me on that theme. Do tell me something about these atoms. I declare it has quite excited me; 'specially because it seems to have something to do with the atomic theory of Dalton.

Hermogenes.—Higgins, if you please. But the thing, as you say, is as old as Democritus, or perhaps older; for Leucippus, Democritus's master, is said to have taught it to him. Nay, there is an older authority still, in the personage (as near to an abstraction as a traditional human being can be) Moschus (not he of the Idyls). But the fact is that the notion of THE ATOM—the *indivisible*, the *thing* that has *place*, *being*, and *power*—is an absolute necessity of the human thinking mind, and is of all ages and nations. It underlies all our notions of being, and starts up, *per se*, whenever we come to look closely at the intimate objective nature of things, as much as space and time do in the subjective. You have dabbled in German metaphysics, and know the distinction I refer to.

Hermione.—You don't mean to say that we are nothing but ATOMS?—Place! being! power! Why, that is I, it is you, it is all of us. Nay, nay. This is going too fast.

Hermogenes.—Perhaps it is.—(You have forgot thought, by-the-by, and will.)—But I am not going to make a single hop quite so far. We shall divide that into two or three jumps, and loiter a little in the intermediate resting-places. But, to go back to your atoms and a vacuum. What does a vacuum mean?

Hermione.—Vacuum? Why, emptiness, to be sure! I mean empty space. Space where *no thing* is. I am not so very sure that I can realize that notion. It is like the abstract idea of a lord mayor that Pope and Atterbury talk about; and in getting rid of the *man*, the gold chain and the custard are apt to start up and vindicate their claim to a place in the world of ideas. And yet I do mean something by empty space. I mean *dis-*

tance—I mean *direction:* that steeple is a mile off, and not *here* where we sit; and it lies south-east of us, and not north or west. And if the steeple were away, I should have just as clear a notion of its *place* as if I saw it there. There now! But then distance and direction imply two *places*. So there are three things anyhow that belong to a vacuum; and let me tell you, it is not everything that three things positively intelligible can be "predicated" of (to speak your jargon).

Hermogenes.—Dear me, Hermione! how can you twit me so? Jargon! Every speciality has its "jargon." Even the Law, that system of dreams, has its "jargon"—the more so, to be sure, because it *is* a system of dreams, or rather of nightmares (God forgive me for saying so!). Well, then, you seem to have tolerably clear notions about a vacuum—at least, I cannot make them clearer. Much clearer, an yhow, than Des Cartes had, who maintained that if it were not for the foot-rule between them, the two ends of it would be in the same place. Still, there is much to be said about that same *Vacuum*, especially when contrasted with a *Plenum*, which means (if it mean anything) the *exact opposite of a vacuum.* In other words, a "jam," a "block," a "fix." But, on the whole, I lean to a vacuum. The other idea is oppressive. It does not allow one to breathe. There is no elbow-room. It seems to realize the notion of that great human squeeze in which we should be landed after a hundred generations of unrestrained propagation.* One does not

* For the benefit of those who discuss the subjects of Population, War, Pestilence, Famine, &c., it may be as well to mention that the

understand how anything could get out of the way of anything else.

Hermione.—Do come back to our dear atoms. I love these atoms: the delicate little creatures! There is something so fanciful, so fairy-like about them.

Hermogenes.—Well they have their idiosyncrasies. I mean, they obey the laws of their being. They comport themselves according to their primary constitution. They conform to the fixed rule implanted in them in the instant of their creation. They act and react on each other according to the rigorously exact, mathematically determinate relations laid down for them *ab initio.* They work out the preconceived scheme of the universe by their—their——col——

Hermione.—Their? Stop, stop! my dear Hermogenes. Where will you land us? Obey laws! Do they know them? Can they remember them? How else can they obey them? Comport themselves according to their primary constitution! Well, that is so far intelligible: they are as they are, and not as they are not. Conform to a fixed rule! But then they must be able to apply the rule as the case arises. Act and react according to

number of human beings living at the end of the hundredth generation, commencing from a single pair, doubling at each generation (say in thirty years), and allowing for each man, woman, and child an average space of four feet in height, and one foot square, would form a vertical column, having for its base the whole surface of the earth and sea spread out into a plane, and for its height 3674 times the sun's distance from the earth! The number of human *strata* thus piled one on the other would amount to 460,790,000,000,000.

determinate relations! I suppose you mean relations with each other. But how are they to know those relations? Here is your atom A, there is your atom B (I speak as you have taught me to speak), and a long interval between them, and no link of connexion. How is A to know where B is; or in what relation it stands to B? Poor dear atoms! I pity them.

Hermogenes.—You may spare your sympathy. They are absolutely blind and passive.

Hermione.—Blind and passive! The more the wonder how they come to perceive those same relations you talk about, and how they "comport themselves," as you call it (*act*, as I should say), on that perception. I have a better theory of the universe.

Hermogenes.—Tell it me.

Hermione.—In the beginning was the nebulous matter, or *Akasch.* Its boundless and tumultuous waves heaved in chaotic wildness, and all was oxygen, and hydrogen, and electricity. Such a state of things could not possibly continue; and as it could not possibly be worse, alteration was here synonymous with improvement. Then came——

Hermogenes.—Now it is my turn to say, Stop! stop! *Solvuntur risu tabulæ.* Do let us be serious. Remember, it was you who began the conversation. *Je me suis seulement laissé entrainer.* The fact is, I have only so far been trying you, and I see you are apt. There lies the real difficulty about these atoms. These same "relations" in which they stand to one another are anything but simple ones. They involve all the "ologies" and all the "ometries," and in these days we know something of what that implies. Their movements, their interchanges, their "hates and loves," their "attractions and repulsions," their "correlations," their what not, are all determined on the very instant. There is no hesitation, no blundering, no trial and error. A problem of dynamics which would drive Lagrange mad, is solved *instanter*, "*Solvitur ambulando.*" A differential equation which, algebraically written out, would belt the earth, is integrated in an eye-twinkle; and all the numerical calculation worked out in a way to frighten Zerah Colburn, George Bidder, or Jedediah Buxton. In short, these atoms are most wonderful little creatures.

Hermione.—Wonderful indeed! Anyhow, they must have not only good memories, but astonishing presence of mind, to be always ready to act, and always *to* act without mistake, according to "the primary laws of their being," in every complication that occurs.

Hermogenes.—Thou hast said it! This is just the point I knew you must come to. The *presence of* MIND is what solves the whole difficulty; so far, at least, as it brings it within the sphere of our own consciousness, and into conformity with our own experience of *what action is.* We know nothing but as it is conceivable to us from our own mental and bodily experience and consciousness. When we know we act, we are also conscious of will and effort; and action without will and effort is to us, constituted as we are, unrealizable, unknowable, inconceivable.

Hermione.—That will do. My head begins to turn round. But I hardly fancied we had got on such an interesting train. We will talk of this again. More to-morrow. Now to the feast of flowers the children are preparing.

COLLINGWOOD, *Aug.* 16, 1860.

XII.

ON THE ORIGIN OF FORCE.*

"Mens agitat molem et magno se corpore miscet."

WHAT is it that we ought to understand by the theory of any natural phænomenon? This is a question not without its importance when we are told, as we so frequently are, that it is useless to inquire into causes: that, in fact, causes are to us as though they were not; seeing that all we can ever attain to is the observation and registry of constant laws of phænomenal sequence:—in other words, that phænomenon succeeds phænomenon, event event, according to certain *rules*, which are all we have any business to inquire into.

(2.) It is unfortunate for this doctrine that within the range of every individual's momentary experience there occurs the phænomenon of *volition;* and that there are

* From the *Fortnightly Review.*

large classes of phænomena, and those most important ones, which, we are quite sure, take place in virtue of such volitions, and without which we are equally sure they would not take place at all. In that peculiar *mental sensation*, clear to the apprehension of every one who has ever performed a voluntary act, which is present at the instant when the determination to do a thing is carried out into the act of doing it—(a sensation which, in default of a term more specifically appropriated to it, we may call that of *effort*)—we have a consciousness of immediate and personal causation which cannot be disputed or ignored. And when we see the same kind of act performed by another, we never hesitate in assuming for him that consciousness which we recognize in ourselves: and in this case we can verify our conclusion by oral communication. The first step in the way of generalization thus taken, the next is obvious enough. Though a flight rather than a step, it forces itself on our thoughts with ever-increasing cogency, the more it is dwelt upon, and the more utterly abortive all attempt to render any other account of that deep mystery of nature—mechanical force—is found to be. Whenever, in the material world, what we call a phænomenon or an event takes place, we either find it resolvable ultimately into some change of place or of movement in material substance, or we endeavour to trace it up to some such change; and only when successful in such endeavour we consider that we have arrived at its theory. In every such change we recognize the action of FORCE. And in the only case in which we are admitted into any personal knowledge of the origin of force, we find it connected (possibly by intermediate links untraceable by our faculties, but yet indisputably *connected*) with volition, and by inevitable consequence, with *motive*, with *intellect*, and with all those attributes of mind in which—and not in the possession of arms, legs, brains, and viscera—personality consists. In limiting thus the domain of physical theory, we keep on the outside of the apparently interminable discussions and difficulties as to the origin of the will itself, which seem to have culminated in some minds in the denial of volition as a matter of fact, and in the dictum of Judge Carleton,* that what men term *the will*, is "simply a passive capacity to receive pleasure from whatever affects us agreeably at the time."

(3.) It may, however, be said, and indeed there are not wanting those who appear very much disposed to say, if not *totidem verbis*, at least by strong implication, that the conception of *Force* itself, as part and parcel of the system of the material universe, is superfluous and therefore illogical. They argue thus. All we know of material phænomena, it is true, resolves itself into the transference of motion from matter to matter. This, however, may be effected by mere collision. Now, when A strikes B, and motion is thereby communicated from A to B, why not at once *admit this as a sequence?* Why interpose an unknown agent, or intermedium, Force, as part of the process? Having come to regard Heat, Light, Electricity, and the "imponderables" generally,

* "Proceedings of the American Philosophical Society," ix. p. 136—Report of Meeting of January 2, 1863.

some upon more, others on less, cogent evidence, as "modes of motion," they seem to consider Force itself as included in the same category, and think there is "reason to believe that it depends on the diffusion of highly-attenuated matter through space."* This doctrine goes to resolve the entire assemblage of natural phænomena into the mere knocking about of an inconceivable number of inconceivably minute billiard balls (or cubes, or tetrahedrons, if that be preferred), which once set in motion and abandoned to their mutual encounter and impact, work out the totality of natural phænomena. With the amount of forethought and intelligence called for in the initiatory disposal of the place and movement of every individual of this multitude; to work out for countless ages the orderly sequence of observed facts, by their blind conformity to the laws of collision, those disposed to adopt such a view of nature would probably concern themselves little. Their actual disposal at the present moment is a *fait accompli;* and from this point it would be possible, at least in imagination, if we knew the present position and movement of every particle of matter in the universe, to work backwards, up to any epoch which we might choose to assign as that of creation, by a simple reversal of the velocity and direction of each:—nay, having thus *un*created the world, the molecules would of themselves work out a pre-existent order of things into all past eternity: an image of what *might have* existed in the past (though it did not) seen, as it were, reflected in the future.

* *North British Review,* vol. xl. p. 41.

(4.) This simplification (if such it be) of our view of material action is altogether untenable; nor will it be difficult, I think (and certainly not superfluous), to show that such an arrangement must of necessity be rapidly self-destructive, and must result in the gradual but speedy dying away of all relative motion, and the reduction of the universe either to a single block of matter moving uniformly on, for ever, in one direction, without relative motion of its parts; or else in the dispersion into space, and absolute final dissociation of its molecules.

(5.) For, be it observed, force (except in the sense of bodily extrusion) being non-existent, our billiard-balls must of necessity be supposed *inelastic.* Elasticity implies force. If this be disallowed,—if elasticity be not *force,* but *collision,*—each billiard-ball (each ultimate atom, that is to say) must be itself a universe in miniature composed of other more minute ones, moving and colliding, *inter se* to give them that resilience which we term elasticity, but which, in this view of the matter, is nothing but "clash." Now what is to prevent these ultimate atoms of the second order, animated with velocities immense, as compared with their mutual distances, *coerced by no mutual attractions*, subject to no control but from their mutual collisions; from dispersing themselves out in all directions into space and abdicating their functions as a group? If we waive this objection (which, however, is fatal) nothing is gained. The original objection applies in its full force to these sub-atoms, and so on *ad infinitum.*

(6.) Now, in the collision of inelastic bodies, *vis viva*

is necessarily and invariably destroyed. The destruction may be total, or may fall short of totality in any proportion according to the directness of the impact, and the proportion of the moving masses; but whenever contact occurs between such bodies, *vis viva* disappears, and, once lost, is gone for ever. Taking such a system in its entirety (*where force exists not*), there is no possibility of its reproduction. There is therefore a necessary and unceasing drain on the *vis viva* of such a system. Everything which constitutes an event, whatever its nature, exhausts some portion of the original stock. Such a system has no vitality. It feeds upon itself, and has no restorative power. All relative motion in it tends rapidly to decay, or at all events to a final state, when there will occur no more collision, *i.e.*, when phænomena cease altogether; when the *minimum* of *vis viva* consistent with the conservation of momentum is attained; and nothing remains but either a single *caput mortuum*, journeying through space, or a multitude of such, travelling different ways; having parted company never to meet again.

(7.) It will of course be urged that this reasoning takes for granted the law just mentioned of the conservation of momentum estimated in any given direction: since we cannot assert *à priori* that two inelastic bodies, after collision, *must* move on with a common velocity and unchanged joint momentum. Of course it does so. But the object of the hypothesis we are combating is to exhibit collision as *a substitute for force; i.e.*, to give an account of the acknowledged laws of motion without introducing the conception of force. We are therefore justified, when arguing against it, in assuming all the results of those laws as established truths: they being, in effect, the very things which the hypothesis is framed to account for. The law in question, constituted as the material universe is, is absolute and universal: and no view of matter and motion can be a true one which is incompatible with it.

(8.) The inward pressure of an etherial medium surrounding the sun upon the earth and planets, suggested by Newton as a mode of escape from the metaphysical difficulty of attraction at a distance, is either only another form of the collision theory above combated, or an evasion of the difficulty by substituting repulsive for attractive force. If the ether press by its elasticity, besides supposing its particles endowed with the necessary amount of repulsion; what, it must be asked, but a repulsion emanating from the sun (and thereby equilibrating and rendering ineffective its inward pressure) is to keep it from rushing in on all sides, and destroying that inequality of density on which its supposed inward pressure depends? If not, its agency must be simply that of an inert resisting medium, rendering the continued revolution of the planets round the sun impossible, and *causing them, while it lasts, to rotate on their axes in a direction contrary to that of their orbital motion*, as a direct consequence of the more rapid abstraction of motion from their outer than from their inner hemispheres. The hypothesis of Le Sage which assumes that *every point of space* is penetrated at *every instant of time* by material particles, *sui generis*, moving in right lines *in every possible direction*, and impinging upon the material atoms of bodies; as a mode of accounting for gravitation, is too grotesque to need serious consideration; and besides, will render no account of the phænomenon of elasticity. Besides this, I am not aware of any other attempt to embody in a tangible form the notion of a substitute for the conception of dynamical force arising out of the elementary conceptions of motion and inertia. There is a tendency indeed, of late apparent, to attribute the elastic pressure of a gas on its containing envelope, as due to the collisive shock of its particles conceived as existing in a continual state of vibration, or of circulation round each other. But the maintenance of such vibrations or revolutions involves the supposition of inter-molecular coercive forces, and is not, therefore, to be classed with such attempts.

(9.) If it be true, then, that the conception of FORCE as the originator of motion in matter without bodily contact, or the intervention of any intermedium, is essential to a right interpretation of physical phænomena; and if it be equally so, on the other hand, that its exertion makes itself manifest to our personal consciousness by that peculiar sensation of *effort* which is not without its analogue in purely intellectual acts of the mind; it comes, not unnaturally, to be regarded as affording a point of contact, a connecting link between these two great departments of being—between mind and matter—the one as its originator, the other as its recipient. The control we possess over the external world we are sure must arise from a capacity somehow inherent in the intellectual part of our nature, to originate or call into action this one and only agent which matter obeys in its changes of form and situation. We may hesitate about admitting into the system of created things around us so vast an amount of additional or extraneous *vis viva*, as the totality of animal exertion since the first introduction of life upon earth would seem to imply. But this is not necessary. The actual *force* necessary to be *originated* to give rise to the utmost imaginable exertion of animal power in any case, may be no greater than is required to remove a single material molecule from its place through a space inconceivably minute—no more in comparison with the dynamical force *disengaged*, directly or indirectly, by the act, than the pull of a hair trigger in comparison with the force of the mine which it explodes. But without the power to make *some* material disposition, to originate *some* movement, or to change, at least temporarily, the amount of dynamical force appropriate to some one or more material molecules, the mechanical results of human or animal volition are inconceivable. It matters not that we are ignorant of the mode in which this is performed. It suffices to bring the origination of dynamical power, to however small an extent, within the domain of acknowledged personality.

(10.) It will perhaps be objected to this, that the principle so generally cited, and now so universally recognized as a dominant one in physics—that of the "conservation of force"—stands opposed to any, even the smallest amount of arbitrary change in the total of "force" existing in the universe. This principle, so far as it rests upon any scientific basis as a legitimate conclusion from dynamical laws, is no other than the well-known dynamical theorem of the conservation of *vis viva* (or of "energy," as some prefer to call it) *supplemented to save the truth of its verbal enunciation*, by the introduction of what is called "potential energy," a phrase which I cannot help regarding as unfortunate, inasmuch as it goes to substitute a truism for the announcement of a great dynamical fact. No such conservation, in the sense of an identity of total amount of *vis viva* at all times, and in all circumstances, in fact, exists. So far as a system is maintained by the mutual actions and reactions of its constituent elements *at a distance* (*i.e.*, by force), *vis viva* may temporarily disappear, and be subsequently reproduced between certain limits. Collision, indeed, between its ultimate atoms, regarded as absolutely rigid, and therefore inelastic (*for that which cannot change its figure can have no resilience*), cannot take place without producing a permanent destruction of it, which there exists no means of repairing. And here we may remark that, this being the case, to ascribe to such atoms any magnitude becomes not only superfluous, but embarrassing. The system of Boscovich has to be accepted in its integrity, if absolute permanence is to be one of the conditions insisted on; and they come to be considered as mere localizations of inertia and such other attributes, including the centralization of force—if any other than this there be—which belong to our notion of material substance. The conservation of energy, then, is in effect no conservation at all in any strict sense of the term, unless so supplemented. It is a fact dynamically demonstratable that the total amount of *vis viva* in any moving system abandoned to the mutual reaction of its particles, while depending at every instant of time, solely for its magnitude, on the then relative situation of those particles (or being, in algebraical phrase, a function of their mutual distances), has a maximum value which it cannot exceed, and a minimum below which it cannot descend. Let its state then be what it will, there is sure to be a certain amount of *vis viva* by which its *actual* falls short of its *extreme possible* value; and to say that the amount of this deficiency added to the actual present amount will make up the maximum, is neither more nor less than a truism: whether expressed in so many words, or by saying that the potential together with the actual energy of the system is invariable; or, again, in other words, that when certain changes have taken place in the relative situations of the parts of the system, what it has lost in actual it has gained in potential energy. When in speaking of a mechanical combination we say that what is lost in time is gained in power, though equally a translation in ordinary language of a dynamical equation, the terms used refer to different modes of viewing the expenditure of force. But in the case before us they stand in their nakedness of similar meaning and convey to the mind no equivalence available for any purpose of reasoning. If, indeed, we could be assured, *à priori*, that the system is one of simple or compound periodicity in which a certain lapse of time will restore every molecule to identically the same relative situation with respect to all the rest; we should then be sure that in the nature of things there would take place periodically, so to speak, a winding up from a lower to a higher state of potential energy, to be subsequently exchanged for newly-created *vis viva*. But as we can have no such *à priori* assurance, can only assume such restoration to be possible, and can see no means of effecting it, if possible, otherwise than by foresight and prearrangement; the one equally with the other is an unknown function, variable within unknown limits, and susceptible of fluctuations to an unknown extent, nor can we have any, the smallest, right to assert that what is expended in the one form *is*, necessarily, laid up in reserve for further use in the other. It would be very diffcult, I apprehend, to show whether in the winding up of a clock, or the building of a pyramid, taking into consideration all the various modes in which *vis viva* disappears and reappears in the expenditure of muscular power, the evolution of animal heat, the consumption of the materials of our tissues (laying aside all question as to the evolution of force from intellectual effort), the propagation of vibratory motion, and a thousand other modes of transfer; the total *vis viva* of this our planet is increased or diminished. That it should remain absolutely unchanged during the process is in the last degree inconceivable. The amount of *vis viva* latent in the form of heat or molecular motion in the sun and planets in our immediate system may bear, and probably does bear, a by no means inappreciable ratio to that more distinctly patent in the form of bodily motion in the periodical circulation of the planets round the sun, and the sun and planets round their axes. The latter amount fluctuates to and fro according to laws easily calculable; but the former we have no means whatever of computing, and to what extent, or within what limits, it may be variable, we are altogether ignorant.

(11.) In what is here said, it is by no means intended to call in question the validity or to underrate the importance of those remarkable physical investigations which have resulted in exhibiting heat as one of the forms in which *vis viva* reappears in the apparent destruction of motion. That all heat *consists in* molecular tremor (or circulation), and is therefore *accompanied* with the alternate development and disappearance of *vis viva* within a limited space and quantity of matter *according to the dynamical laws of such tremulous or rotating movements*, may very readily be granted. But that there are no forms of internal molecular movement other than heat, and what we now speak of as its "correlated forces" in which *vis viva* may be temporarily stored up, to make its appearance ultimately in a form cognizable to our senses, is what can by no means be so readily admitted. Nor (while accepting with all due admiration as approximate truths these great revelations as to the mutual convertibility of these correlatives according to the measure of *vis viva* appropriate to each) shall we advance any nearer to a rational theory of any one of them, till it shall be shown with much more distinctness than at present appears, in what these molecular movements themselves consist; by what forces (in the dynamical acceptation of the term) they are controlled;

in what manner, or by what mechanism, they are propagated from one body to another; and how their mutual interconversion is effected. In referring them to the action of dynamical force upon matter, and in getting rid of the "imponderables" (other than the luminiferous ether) we are at length fairly entered on the construction of a *theory* of their phænomena, in what, as above remarked, must be considered the true acceptation of that term in physics: and once satisfied that dynamical force itself is a phænomenon *sui generis;* that it is not a result of collision — an *educt* from the duality *Inertia* and *Motion;*—one of those correlatives, in short, to which the epithet "Physical forces" has of late been so generally, and, in my opinion, so very improperly applied, we have reached the point where theory ends and speculation begins,—where we cease to inquire into the *causes* of phænomena, and direct our consideration thenceforward to their *reasons.*

(12.) The universe presents us with an assemblage of phænomena, physical, vital, and intellectual—the connecting link between the worlds of intellect and matter being that of organized vitality, occupying the whole domain of animal and vegetable life, throughout which, in some way inscrutable to us, movements among the molecules of matter are originated of such a character as apparently to bring them under the control of an agency other than physical,* superseding the ordinary laws

* Take for instance the formative *nisus,* which determines the production of a supernumerary finger in the human hand. Here is no gradual change from generation to generation, no first develop-

which regulate the movements of inanimate matter, or, in other words, *giving rise to movements* which would not result from the action of those laws uninterfered with; and therefore implying, on the very same principle, the origination of force. The first and greatest question which Philosophy has to resolve in its attempts to make out a Kosmos,—to bring the whole of the phænomena exhibited in these three domains of existence under the contemplation of the mind as a congruous whole,—is,

ment of a rudimentary joint followed in slow succession after centuries of hereditary improvement, by the others, up to the perfect member. It starts at once into completeness. The change in the *working-plan* of the whole hand has been carried out at once, by a systematic engraftment of blood-vessels and nerves into effective connexions with the centres of nutritive, mechanical, and sensitive action in the frame, as if by some preconceived arrangement. [Since this was written I have been informed of two or three instances of superfluous *thumbs*. They were imperfectly formed, not movable, and so far might be considered rudimentary.]

In direct reference to this point I would call the reader's attention to a very striking passage in the Croonian Lecture for 1865, delivered before the Royal Society by Prof. Beale, where, after stating that "phænomena occur in the simplest form of living matter, which never have been, and which," he believes, "never can be explained upon any known physical or chemical laws"—he goes on to say,—

"Living matter is not a machine, nor does it act upon the principles of a machine, nor is force conditioned in it as it is in a machine, nor have the movements occurring in it been explained by physics, or the changes which take place in its composition by chemistry. The phænomena occurring in living matter are peculiar, differing from any other known phænomena; and therefore, until we can explain them, they may well be distinguished by the term *vital*. Not the slightest step has yet been made towards the production of matter possessing the properties which distinguish living matter from matter in every other known state."—*Proceedings of the Royal Society*, xiv. p. 282, No. 72.

whether we can derive any light from our internal consciousness of thought, reason, power, will, motive, design —or not: whether, that is to say, nature is or is not *more interpretable* by supposing these things (be they what they may) to have had, or to have, to do with its arrangements. Constituted as the human mind is, if nature be *not* interpretable through these conceptions, it is not interpretable at all; and the only reason we can have for troubling ourselves about it is either the utilitarian one of bettering our condition by "subduing nature" to our use through a more complete understanding of its "laws," so as to throw ourselves into its grooves, and thereby reach our ends more readily and effectually; or the satisfaction of that sort of aimless curiosity which can find its gratification in scrutinizing everything and comprehending nothing. But if these attributes of mind are not consentaneous, they are useless in the way of explanation. Will without Motive, Power without Design, Thought opposed to Reason, would be admirable in explaining a chaos, but would render little aid in accounting for anything else.

XIII.

ON THE ABSORPTION OF LIGHT BY COLOURED MEDIA,

VIEWED IN CONNEXION WITH THE UNDULATORY THEORY.*

HE absorption of light by coloured media is a branch of physical optics which has only since a comparatively recent epoch been studied with that degree of attention which its importance merits. The speculations of Newton on the colours of natural bodies, however ingenious and elegant, can hardly, in the present state of our knowledge, be regarded as more than a premature generalization; and they have had the natural effect of such generalizations, when specious in themselves and supported by a weight of authority admitting for the time of no appeal, in repressing curiosity, by rendering further inquiry apparently superfluous, and turning attention

* The substance of this paper was read before the Section of Physics of the British Association, at Cambridge, 1833. Some inaccuracies of wording are corrected, but nothing *introduced* bearing on the views more recently entertained as to the conversion of motion into *heat-vibration.*

into unproductive channels. I have shown, I think satisfactorily, however, in my Article on Light,* that the applicability of the analogy of the colours of thin plates to those of natural bodies is limited to a comparatively narrow range, while the phænomena of absorption, to which I consider the great majority of natural colours to be referrible, have always appeared to me to constitute a branch of photology *sui generis* to be studied in itself by the way of inductive inquiry, and by constant reference to facts as nature offers them.

(2.) The most remarkable feature in this class of facts consists in the unequal absorbability of the several prismatic rays, and the total abandonment of anything like regularity of progress in this respect as we proceed from one end of the spectrum to the other. When we contemplate the subject in this point of view, all idea of regular functional gradation is at an end. We seem to lose sight of the great law of continuity, and to find ourselves involved among desultory and seemingly capricious relations, quite unlike any which occur in other branches of optical science. It is, perhaps, as much owing to this as to anything, that the phænomena of absorption in some recently-published speculations, and in the view which Mr Whewell has taken in his Report of the progress and actual condition of this department of natural philosophy, read to this meeting, have been characterized as peculiarly difficult to reconcile with the undulatory theory of light. In so far as I have above

* This refers to the Article on Light published in the "Encyclopædia Metropolitana" in 1826-7.

described the phænomena in appropriate terms, it will be evident that a certain difficulty must attach to their reduction under the dominion of *any* theory, however competent, ultimately, to render a true account of them. Where such evidence of complication and suddenness of transition subsists on the face of any large assemblage of facts, we are not to expect that the mere mention of a few general propositions, like cabalistic words, shall all at once dissipate the complication, and render the whole plain and intelligible. If we represent the total intensity of light, in any point of a partially-absorbed spectrum, by the ordinate of a curve whose abscissa indicates the place of the ray in order of refrangibility, it will be evident, from the enormous number of maxima and minima it admits, and from the sudden starts and frequent annihilations of its value through considerable amplitudes of its abscissa, that its equation, if reducible at all to analytical expression, must be of a singular and complex nature; and must at all events involve a great number of arbitrary constants dependent on the relation of the medium to light, as well as transcendents of a high and intricate order. We must not, therefore, set it down to the fault of either of the two rival theories if we do not at once perceive how such phænomena are to be reconciled to the one or to the other; but rather endeavour to satisfy ourselves whether there be, in the first instance, anything in the phænomena, generally considered, repugnant either to sound dynamical principles, or to the notions which those theories respectively involve as fundamental features.

(3.) Now, as regards only the general fact of the obstruction and ultimate extinction of light in its passage through gross media, if we compare the corpuscular and undulatory theories, we shall find that the former appeals to our ignorance, the latter to our knowledge, for its explanation of the absorptive phænomena. In attempting to explain the extinction of light, on the corpuscular doctrine, we have to account for the light so extinguished as a material body, which we must not suppose annihilated. It may, however, be transformed; and among the imponderable agents, heat, electricity, &c., it may be that we are to search for the light which has become thus comparatively stagnant. The heating power of the solar rays gives a *primâ facie* plausibility to the idea of a transformation of light into heat by absorption. But when we come to examine the matter more nearly, we find it encumbered on all sides with difficulties. How is it, for instance, that the most luminous rays are not the most calorific, but that, on the contrary, the calorific energy accompanies, in its greatest intensity, rays which possess comparatively feeble illuminating powers? These and other questions of similar nature may perhaps admit of answer in a more advanced stage of our knowledge; but at present there is none obvious. It is not without reason, therefore, that the question, "What becomes of light?" which appears to have been agitated among the photologists of the last century, has been regarded as one of considerable importance as well as obscurity by the corpuscular philosophers.

(4.) On the other hand, the answer to this question

afforded by the undulatory theory of light is simple and distinct. The question, "What becomes of light?" merges in the more general one, "What becomes of motion?" And the answer, on dynamical principles, is, that it continues for ever. No motion is, strictly speaking, annihilated; but it may be divided, and the divided parts made to oppose and, in point of ultimate effect, counteract each other. A body struck, however perfectly elastic, vibrates for a time, and then appears to sink into its original repose. But this apparent rest (even abstracting from the inquiry that part of the motion which may be conveyed away by the ambient air), is nothing else than a state of subdivided and mutually-compensating motion, in which every molecule continues to be agitated by an indefinite multitude of internally-reflected waves, propagated through it in every possible direction, from every point in its surface on which they successively impinge. The superposition of such waves will, it is easily seen, at length operate their mutual counteraction, which will be the more complete, the more irregular the figure of the body and the greater the number of internal reflections.

(5.) In the case of a body perfectly elastic and of a perfectly regular figure, the internal reflection of a wave once propagated within it in some particular direction might go on for ever without producing mutual destruction; and in sonorous bodies of a highly elastic nature we do in fact perceive it to continue for a very long time. But the least deviation from *perfect elasticity* resolves our conception of the vibrating mass into that of a multitude of inharmonious systems communicating with each other.

At every transfer of an undulation from one such system into that adjacent, a partial echo is produced. The unity of the propagated wave is thus broken up, and a portion of it becomes scattered through the interior of the body in dispersed undulations from each such system, as from a centre of divergence. In consequence of the continual repetition of this process, after a greater or less number of passages to and fro of the original wave across the body (however perfect we may suppose the reflections from its surface to be), it becomes frittered away to an insensible amplitude, and resolved into innumerable others; crossing, recrossing, and mutually compensating each other, while each of the secondary waves so produced is in its turn undergoing the same process of disruption and degradation.

(6.) In this account of the apparent destruction of motion, I have purposely supposed the body set in vibration to be insulated from communication with any other. In the case of a perfectly or highly elastic body struck in air, it will vibrate so long that a great part of its motion is actually carried off in sonorous tremors communicated to the air. But in the case of an inelastic or imperfectly elastic body, the internal process above described goes on with such excessive rapidity, as to allow of very few, and those rapidly degrading, impulses to be communicated from its surface to the air.

(7.) In my Article on Sound,* I have explained, on this principle of internal reflection and continual subdivision,

* Published like that on Light, above cited, in the *Encyclopædia Metropolitana*, 1829–30.

in a medium consisting of loosely-aggregated earth intermixed with much air, the hollow sounds which are often attributed to the reverberation of subterrannean cavities, and in particular the celebrated instance of this kind of sound heard at the Solfaterra near Pozzuoli. The dull and ill-defined sound thus produced from a succession of partial echoes is there assimilated to the nebulous light which illuminates a milky medium when a strong beam is intromitted. If we suppose, now, such a mass of materials insulated from communication with the external air by some *sound-tight* envelope, these partial echoes, when they reach the surface in any direction, will be all sent back again as so many fresh impulses, till at length it will become impossible to assign a point within the mass which will not be agitated at one and the same moment by undulations traversing it in every possible phase and direction. Now the state of a molecule, under the influence of an infinite number of contradictory impulses thus superposed, is undistinguishable from a state of rest.

(8.) The only difficulty, then, which remains in the application of the undulatory theory to the absorptive phænomena, is to conceive how a medium (*i.e.*, a combination of æthereal and gross* molecules) can be constituted so as to be transparent, or freely permeable to one ray or system of undulations, and opake, or difficultly per-

* By *gross* molecules, or gross bodies, I understand the *ponderable* constituents of the material world, whether solid, liquid, or gaseous; using the term in contradistinction to æthereal, which has reference to the luminiferous æther.

meable to another, differing but little in frequency. Now it is sufficient for our present purpose if, without pretending to analyze the actual structure of any optical medium, we can indicate structures and combinations in which air, in lieu of the æther, is the undulating medium, and which shall be either incapable of transmitting a musical sound of a given pitch, or shall transmit it much less readily than sounds of any other pitch, even those nearly adjacent to it. For that which experiment, or theory so well grounded as to be equally convincing with experiment, shows to be possible in the case of musical sounds, will hardly be denied to have its analogue or representative among the phænomena of colour, when referred to the vibrations of an æther.

(9.) An example of an acoustic combination, or compound vibrating system, incapable of transmitting a

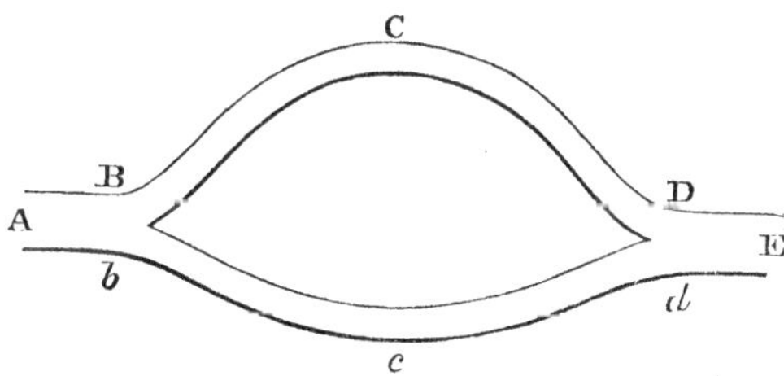

musical sound of a given pitch, is furnished by the pipe A E, which, after proceeding singly a certain length A B, at B branches off into two equal and symmetrically disposed pipes B C and *b c*, which reunite again at D *d*, and there again constitute a single pipe D E, whose direction shall (like A B) bisect the angle between the branches. The branches, however, are of unequal length, the one B C D being longer than the other, by a quantity equal to half the length of the undulation or pulse of the musical note in question. It is evident, then, that if that note be sounded at A, each pulse will subdivide itself at B *b*, and the divided portions will run on along the two branches with equal intensities till they reunite at D *d*. They will arrive there, however, in opposite phases, and will therefore counteract each other at their point of reunion, and in every point of their subsequent course along the pipe D E; so that on applying the ear at E no sound should be heard, or at best a very feeble one, arising from some slight inequality in the intensities wherewith the undulations arrive by the longer and shorter pipe—a difference which may be made to disappear, by giving the longer a trifle larger area for its section.*

(10.) Suppose now that the pipes instead of being cylindrical were square, and that the whole surface of one side of a chamber were occupied with the orifices A of such pipes, leaving only such intervals as might be necessary to give room for their due support, and for

* I ought to observe, that I have not *made* the experiment described in the text, nor am I aware that it has ever been made; but it is easy to see that it ought to succeed, and would furnish an apt enough illustration of the principle of interference. Instead of a pipe, inclosing air, a canal of water might be used, in which waves of a certain breadth, excited by some mechanical contrivance at one end, would not be propagated beyond the point of reunion, D, of the two canals into which the main channel, A B, was divided.

their subdivision according to the condition above explained; and suppose, further, that the other ends (E) of all the reunited pipes opened out, in like manner, into another chamber, at some considerable distance from the first, and separated from it by masonry or some material, filling in all the intervals between the pipes, so as to be completely impervious to sound. Things being so disposed, let the whole scale be sounded, or a concert of music performed in the first chamber, then will every note, except that one to which the pipes are thus rendered impervious, be transmitted. The scale, therefore, so transmitted, will be deficient by that note, which has been, to use the language of photologists, *absorbed* in its passage. If several such chambers were disposed in succession, communicating by compound pipes, rendered impervious (or *un*tuned, as we may term it) to so many different notes, all these would be wanting in the scale on its arrival in the last chamber; thus imitating a spectrum in which several rays have been absorbed in their passage through a coloured medium.

(11.) In my Article on Light, above referred to, Art. 505, I have suggested, as a possible origin of the fixed lines in the solar spectrum, and (*pari ratione*) of the deficient or less bright spaces in the spectra of various flames, that the same indisposition in the molecules of an absorbent body to permit the passage of a particular coloured ray *through* them, may constitute an obstacle, *in limine*, to the production of that ray *from* them. The following easy experiment will explain my meaning. Take two tuning forks of the same pitch, and heating the ends of them, fasten with sealing-wax, on one of them one, and on the other two, disks of card (all equal in size), on the *inner* surfaces, having the plane of the card perpendicular to that of a section of the fork through the axes of both its branches. The cards on that fork which has two, should have their surfaces about a tenth of an inch asunder, and their centres just opposite; and the other fork should be brought into unison with it by loading its undisked branch with additional wax, equal in weight to the disk and wax on the other. Now strike the forks, and a remarkable difference will be perceived in the intensity of their sounds. The fork with one disk will utter a clear and loud sound, while that of the other will be dull and stifled, and hardly audible, unless held close to the ear. The reason of this difference is that the opposite branches of the fork are always in opposite states of motion, and that in consequence the air is agitated by either the two branches vibrating freely, or by both loaded with equal disks, with nearly equal and opposite impulses; whereas in the case of a fork furnished with only one disk, a greater command of the ambient medium is given to the branch carrying it, and a much larger portion of uncounteracted motion is propagated into the air. Here, then, we have a case in which a vibrating system in full activity is rendered, by a peculiarity of structutre, incapable of sending forth its undulations with effect into the surrounding medium; while the very same mass of matter, *vibrating with the same intensity*, but more favourably disposed as to the arrangement of its parts, labours under no such disability.

(12.) The disked tuning fork is a most instructive instrument, and I shall not quit it until I have availed myself of its properties to exemplify the easy propagation of vibrations, of a definite pitch, through a system comparatively much less disposed to transmit those of any other pitch. Take two or more forks in unison, and furnish each of them with a single disk of the size of a large wafer, looking outwards. Having struck one of them, let its disk be brought near to that of the other, centre opposite to centre, and it will immediately set the other in vibration, as will be evident by the sound produced by it when the first fork is stopped, as well as by its tremors, sensible to the hand which holds it. The communication of the vibration is much more powerful and complete when a small loop of fine silver wire is fixed to one of the forks, and brought lightly into contact with the other, with its looped or convex side. Imagine now a series of such forks and loops arranged as in the annexed figure, and let the first, A, be maintained in vibration by any exciting cause, as, for instance, by sounding a musical note opposite to its disk, A, in unison with its pitch. The vibrations so excited will, as is evident, run along the whole line, though with diminishing intensity, to the last

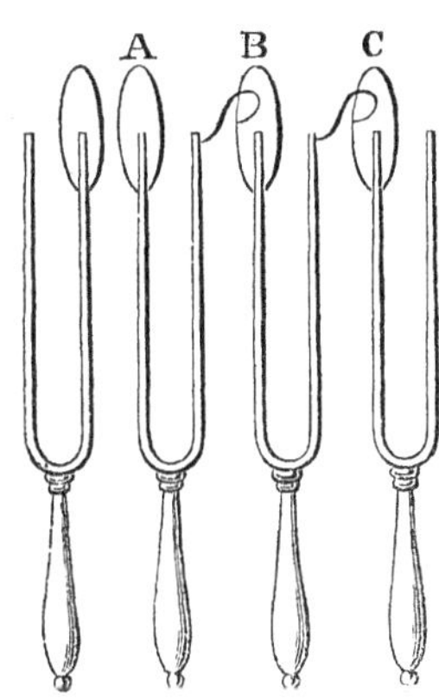

fork. Here, then, we have a case analogous to the easy transmission of a ray of definite colour, accompanied with its gradual extinction, in traversing a considerable thickness of the absorbing medium. If we would avoid the actual contact of the vibrating systems, we may conceive an arrangement like that here depicted, where, in place of forks, straight bars, disked at both ends and supported at their centres, are used to form the vibrating series.

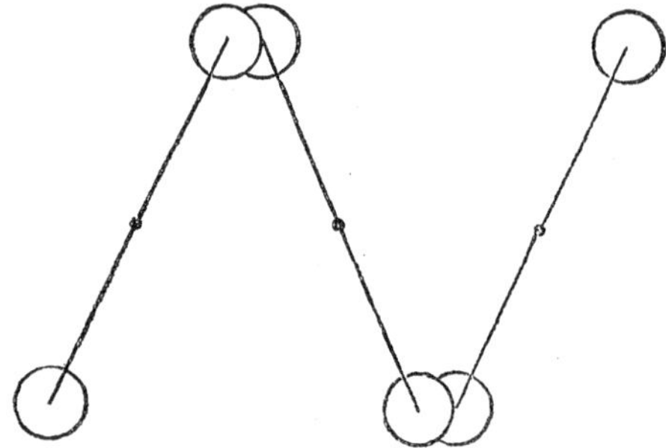

(13.) When two disked tuning forks slightly out of unison are opposed to each other, the vibrations of one are still communicated to the other, even when they differ sufficiently to produce audible and pretty rapid beats. But the communication in this case is less complete, and the sound produced feebler, than in that of perfect unison, and the degradation of intensity in the communicated sound is very rapid as the forks recede from unison. We have here a fact analogous to the appearance of a bright line in the spectrum situated between dark spaces, and as it is not difficult to imagine combinations of the nature above mentioned, in which several different notes shall be transmitted, while the intermediate one, finding no unisons, or near approaches to unison in the systems established, shall be extinguished; so by analogy we may perceive how any number of bright and dark lines may be produced in a spectrum unequally absorbed.

(14.) The case last put is entirely analogous in its principle to that of a phænomenon which is described in my Treatise on Sound, and of which, at the time of the publication of that Treatise, I believed myself to have been the first and only observer, though I have recently learned to rectify that impression, and have great pleasure in referring the experiment, which is a remarkably easy and striking one, to Mr Wheatstone, the author of so many other ingenious and instructive experiments in this department of physics. If a tuning fork be held over the open end of a pipe pitched in unison with it, the pipe will *speak* by resonance. (If the fork be disked, and the aperture of the pipe be nearly covered by the disk, the tone brought out is one of a clearness and purity quite remarkable.) Now both Mr Wheatstone and myself have observed that if *two forks*, purposely pitched out of unison with each other, so as to yield the beats of imperfect concords, be *at once* held over the orifice, the pipe will, at one and the same moment, yield both the notes, and will utter loud beats, being actually out of unison with itself. In proportion, however, as the pitch of one or other fork deviates from that to which the length of the pipe corresponds, and which the pipe alone would utter, the resonance of its tone is feeble, and beyond a certain interval becomes inaudible.

(15.) The dynamical principle on which these and similar phænomena depend is that of "forced vibrations," as it is stated in the Essay on Sound above referred to, or, more generally, in a more recent publication (Cab. Cyclop., volume on Astronomy), in terms as follow: "If one part of any system, connected either by material ties or by the mutual attractions of its members, be continually maintained by any cause, whether inherent in the constitution of the system or external to it, in a state of regular periodic motion, that motion will be propagated throughout the whole system, and will give rise in every member of it, and in every part of each member, to periodic movements, executed in equal periods with that to which they owe their origin, though not necessarily *synchronous* with them in their maxima and minima." The general demonstration of this as a dynamical theorem is given in the Essay on Sound already referred to, and its applicability to the transmission of light through material bodies is indicated in a note thereto appended.

(16.) The mode, then, in which we may conceive the transmission of light through gross media to be performed, so as to bring the absorptive phænomena within the wording of this principle, is, to regard such media as consisting of innumerable distinct vibrating parcels of molecules, each of which parcels, with the portion of the luminiferous æther included within it (with which it is connected, perhaps, by some ties of a more intimate nature than mere juxtaposition), constitute a distinct compound vibrating system, in which parts differently elastic are intimately united and made to influence each other's motions. Of such systems in acoustics we have no want of examples—in membranes stretched on rigid frames, in cavities stuffed with fibrous or pulverulent substances, in mixed gases, or in systems of elastic laminæ, such as boards, sheets of glass, reeds, tuning forks, &c., each having a distinct pitch of its own, and all connected by some common bond of union. In all such systems the whole will be maintained in forced vibration so long as the exciting cause continues in action, but the several constituents, regarded separately, will assume, under that influence, widely different amplitudes of oscillation, those assuming the greatest whose pitch taken singly is nearest to coincidence with that of the exciting vibrations. Everybody is familiar with the tremor which some particular board in a floor will assume at the sound of some particular note of an organ; but when that note is not sounded, it is sufficiently apparent that the board is no less occupied in performing its dynamical office of transmitting to the soil below, or dispersing through its own substance and the contiguous bodies, the motion which the oscillation of the air above is continually imparting to it.

(17.) As we know nothing of the actual forms and intimate nature of the gross molecules of material bodies, it is open to us to assume the existence, in one and the same medium, of any variety of them which may suit the explanation of phænomena. There is no necessity to suppose the luminiferous molecules of gross bodies to be identical with their ultimate chemical atoms. I should rather incline to consider them as minute groups, each composed of innumerable such atoms; and it may be that in what are called uncrystallized media, the axes or lines of symmetry of these groups may have no particular direction, or rather all possible directions, or the groups themselves may be unsymmetrical. Such a disposition of things would correspond with a uniform law of absorption, independent of the direction of the transmitted ray; while in crystallized media a uniformity of constitution and position of these elementary groups, or rather of the cells or other combinations which they may be regarded as forming with the interfused æther, may be readily supposed to draw with it differences in their mode of vibration, and even different disposals of their nodal lines and surfaces, according to the different directions in which undulations may traverse them: and which may not impossibly be found to render an account of the change of tint of such media according to the direction of the rays in their interior, as well as of the different tints and intensities of their oppositely polarized pencils; of which latter class of phænomena, however, I shall immediately have occasion to speak further.

(18.) But as my present object is merely to throw out, as a subject for examination, a hint of a possible explanation of the phænomena of absorption, on the undulatory theory, I shall not now pursue its application into any detail, nor attempt the further development of particular laws of structure competent to apply to this or that phænomenon. I will, however, mention one or two facts in acoustics which appear to me strongly illustrative of corresponding phænomena in the propagation of light. The first of these is the impeded propagation of sound in a mixture of gases differing much in elasticity as compared with their density. The late Sir J. Leslie's experiments on the transmission of sound through mixtures of hydrogen with atmospheric air sufficiently establish this remarkable effect. It would be desirable to prosecute those experiments in larger detail, but hitherto I am not aware of anybody having ever repeated them. It would be interesting, for instance, to inquire whether the impediment offered by such a mixture of gases be the same for all *pitches* of a musical note, or not; and how far this phænomenon might be imitated by mixing actual *dust* of a uniform size of particle, such as the dust of Lycoperdon, &c.; or aqueous fog, and how far such mixture would affect unequally sounds of different pitches.

(19.) The other fact in the science of acoustics which I would notice as illustrative of a corresponding phænomenon in photology, is one observed by Mr Wheatstone, which I have his permission to mention. In attempting to propagate vibrations along wires, rods, &c., to great distances, he was led to remark a very great difference in respect of facility of propagation between vibrations longitudinal and transverse to the general direction of propagation. The former were readily conveyed with almost undiminished intensity to any distance; the latter were carried off so rapidly by the air, as to be incapable of being transmitted with any considerable intensity to even moderate distances. This strikes me as obviously analogous to the ready transmissibility of a ray polarized in one certain direction, through a tourmaline or other absorbing doubly-refracting crystal, while the oppositely-polarized ray (whose vibrations are rectangular to those of the first) is rapidly absorbed and stifled, *i.e.*, dispersed, by the agency of the colouring matter which acts the part of the air in Mr Wheatstone's experiment, and self-neutralized by the opposition of its subdivided portions as above explained.

SLOUGH, *October* 19, 1833

XIV.

ON THE ESTIMATION OF SKILL IN TARGET-SHOOTING.

HAPPENING to be present at an archery meeting at St Leonard's not long ago, I was struck, on examining the target, by observing a much less degree of concentration of "hits," as indicated by the marks left on them about the centre and within the "gold" (accompanied by a rapidly-increasing frequency of marks within the immediately-surrounding circles, proceeding thence outwards from the centre), than could be at all reconciled with my own preconceived impression of what ought to be the proportional numbers, according to what I then considered as results afforded by a legitimate application of the calculus of probabilities to such a question. The proportional numbers of hits in the several areas distinguished as gold, red, blue, white, black, marked out by equidistant rings of 4·8 inches each in breadth, surrounding the gold circle of that radius, ought, according to the statement given by myself in my review of M. Quetelet's work on Probabilities,* to run as follows:—In the gold (out of 500 hits), 107; in the red annulus, 106; in the blue, 101; in the black, 97; and in the white, 89; supposing the target (terminating with the white) to receive half the entire number (1000) of arrows discharged; which in the case observed was not far from the truth. Whereas by the actual record of that day's shooting, handed to me afterwards, the proportional numbers corresponding to a total of 500 hits were:—Gold, 31; red, 89; blue, 121; black, 140; white, 119. This discordance with observation, being far too great to be attributable to ordinary casualty (the whole number of arrows discharged on the day in question being upwards of 7000), led me, of course, to re-examine the reasoning on which the first expectation had been grounded. And so enlightened, I was at no loss to discover its fallacy,—affording, as it does, a good example of the necessity of close attention to the wording of all reasonings on questions of probability. It was, in fact, traceable to the wording of a proposition perfectly true, and, as applied to the case where it was employed in another inquiry, correctly applicable, viz.,† "Suppose a ball dropped from a given height, with the intention that it shall fall on a given mark. Fall as it may, its deviation from the mark *is error;* and the probability of that error . . . . decreases in geometrical

* Essays from the Edinburgh and Quarterly Reviews, &c., &c. Longman, 1857. P. 401.

† Essays, &c., &c. Pp. 398, 399.

progression as the square of the error increases in arithmetical." Now, it is perfectly true that the deviation of the point of incidence from the mark *is error*. But it is something more special. It is error *in that one particular direction* in which the point of incidence lies from the mark aimed at. In estimating, therefore, the probability of striking a target *at a certain definite distance* from the centre aimed at, we must multiply the probability of striking *a determinate point* at that distance from the centre, by the number of points within the extent of the target which actually do lie at that distance from it, without regard to the directions in which they lie: *i.e.*, we have to multiply the fractional number expressing the abstract probability of committing a given error out of an indefinite number of *equally possible* ones, by a number proportional to the *degree of opportunity* which the circumstances of the special case afford for its commission. In this case that degree of opportunity is evidently measured by, and proportional to, the length of the circumference of the circle about whose centre, at the distance specified, an arrow may strike, or a ball drop from a height.

(2.) Reasoning on this (the correct principle in the case of target-shooting), any one conversant with mathematical analysis will find no difficulty in arriving at the following singularly neat and simple formula for the *probability of missing*, in any one single shot, a circular area of given radius (r), at whose centre the shooter aims.*

* The demonstration of this formula is annexed in the form of a note at the end of this essay.

Denoting by a the radius of the circular area within which his skill would, on the average of an immense number of shots, enable him to plant half the total number discharged; and by M the fraction expressing the probability in question, certainty being expressed by 1, we shall have

$$M = \left(\frac{1}{2}\right)^{\frac{r^2}{a^2}}$$

while for H the *probability of hitting* the same area we have

$$H = 1 - M$$

(3.) From these expressions, knowing the value of a, which is the inverse measure of the skill of the shooter (being less the greater that skill), it is easy to calculate his chance of hitting a circle of any given radius in a single shot. And, reversing the question, his skill (measured by the fraction $\frac{1}{a}$) may be ascertained, by observing what percentage of shots he can plant, on a large average, from a given distance, within a circle of any given radius (r). For that percentage being the numerical expression of his probability of hitting the circle, or the value of H, or $1-M$, M is known, and a will be given by the formula.

$$a = r. \sqrt{-\frac{Log.\ 2}{Log.\ M.}} = r. \sqrt{-\frac{Log.\ 2}{Log.(1-H)}}$$

Thus, if a marksman be observed to plant 9 per cent. of his arrows within a circle of one foot in diameter at the distance of one hundred yards, we have

$r = \overset{\text{f.}}{\tfrac{1}{2}}$; $H = \frac{9}{100}$; $M = \frac{91}{100}$, whose logarithm is —0·04096, that of 2 being + 0·30103: so that $a = \frac{1}{2}\sqrt{\frac{30103}{4096}}$ $= \overset{\text{f.}}{1\cdot355} = \overset{\text{f. in.}}{1\cdot4\tfrac{3}{4}}$; which, doubled, gives $\overset{\text{f. in.}}{2\cdot8\tfrac{1}{2}}$ for the diameter of a target which he might make an even bet to hit at the first shot. And according to the values of this constant, so determined in the case of each several competitor, ought their names to be arranged in a prize-list, the smaller values ranking higher than the larger.

(4.) If the object of the competition be merely to arrange the competitors correctly in order of skill at the moment, without deducing for each any definite and normal numerical result expressive of his absolute skill, and comparable with others derived from practice with targets of other dimensions, and at other distances; it is evident that the trouble of any such computation as the above may be spared, since the same precise *order* must necessarily result from merely tabulating the total number of hits of each competitor (practising with an equal number of arrows, and at one and the same distances). Were the number of shots allowed to each immensely large, the same order of merit and the same set of values of the constant a would result from a record of the hits *within the total area* of each of the several circles marked out by the outer circumferences of the gold, red, blue, black, and white colours. The only use of these rings is to give opportunity for a variety of prizes, and that piquancy and interest to the result of a day's shooting which arises from the element of *luck* mixing itself in the competition. This it does the more, the fewer the shots allowed to each, nor can it be eliminated, so as to make skill the sole determining power, but on the average of very enormous numbers, such as, for instance, ten or twenty thousand arrows discharged by each marksman. Every shooter, of course, *aims* to the best of his ability, exclusively with a view to hit the centre of the gold; nor is it conceivable that, having that intention, there should exist in any individual such specialty of aiming as should disperse his shots, failing the gold, so as to strike preferentially (say) the blue, rather than the red ring on one side of it and the black on the other.

(5.) The following table shows the respective numbers of "hits" per 1000 shots, which may be expected to occur on a calculation from our formulæ, within the several coloured areas of the five equidistant rings (considering the central gold as the first ring) into which an ordinary target is divided. Considering its diameter as divided into ten equal parts, the outside diameters of those rings will be respectively 2, 4, 6, 8, 10; their radii, 1, 2, 3, 4, 5; and the areas of their containing circles, in the proportion of the squares of these numbers, 1, 4, 9, 16, 25,—so that the areas of the several coloured spaces form the progression 1, 3, 5, 7, 9. The usual rule of valuation, then, which accords to hits in any of the rings (from the white inwards), values in the proportion of these numbers; assumes the probability of hitting to be in the simple proportion of the area struck (as would be the case were the shooting entirely at random), and the merit to increase as the probability, *so estimated* diminishes. The range in this table of the quantity a, or what may be termed the probable error from the centre of a single shot, includes what may be taken as the extremes of good and bad shooting :—

TABLE I.

a	Probable hits per thousand in the					Misses 5 to ∞.
	Gold 0 — 1.	Red 1 — 2.	Blue 2 — 3.	Black 3 — 4.	White 4 — 5.	
1	500	438	42	20	0	0
2	159	341	290	147	50	13
3	74	191	235	208	146	146
4	42	117	164	177	161	339
5	27	78	116	137	142	500
6	19	55	85	106	118	617
7	14	41	65	82	96	702
8	11	31	51	66	78	763
9	9	25	40	54	65	807
10	7	20	33	45	54	841

(6.) For the purpose of comparing this theory with practice, I have been favoured with the series of annual reports of the practice at the Grand National Archery Meeting, with their target lists, and awards of prizes, for fifteen successive years, commencing with 1850; which record the hits made by each competitor in each of the colours, from specified distances, and with specified numbers of arrows. The number of shots delivered amounts, collectively, to upwards of half a million; and excluding 169 cases in which it is noted that the shooter did not deliver all his arrows, and those comparatively much more rare ones in which the record of the shots is incompatible with the awarded value from some other cause than a mere misprint (which can generally be rectified), to 474,384; of which 168,239 were hits, and 306,145 misses, on a target of 48 inches in diameter, which gives 4·8 in. for the unit of our a. This gives as a general average, 644 misses per 1000 shots; and therefore were the distances all alike, or for an average distance of 80 yards, would correspond to a value of a in our table, of very nearly $6\frac{1}{3}$, or to a mean probable deviation of a single shot, of 30·4 in.; the total number of competitors being 2075 (reckoning the same individual appearing in several lists as so many distinct competitors), shooting at 430 targets. As the causes of linear deviation may be considered as increasing proportionally to the distance, and as in fixing the average distance as above at 80 yards, (strictly 79·1,) the number of arrows discharged at each distance is taken into account; this may be regarded as a fair estimate of our national proficiency in archery, and as comparable, in the terms of its statement, with what may be obtained at a future period, or in other countries.

(7.) In deducing the results embodied in the following tables, the numbers of hits made by all the shooters at each target in its several colours were summed separately. The results so obtained for all the targets for each class of shooters, and for each distance, in each year of the series, were then grouped together and summed, and the fifteen sets of annual sums so obtained, united into general sums, as exhibited below in Table II., to shorten which it is to be borne in mind that of the lady competitors, 853 in number, each delivered 96 arrows at 60 yards, and 48 at 50;* and the

* In the year 1850 the arrows delivered by each lady were only 72 and 36 at the same distances.

gentlemen, numbering 1222, 144 arrows at 100 yards, 96 at 80, and 48 at 60 yards: the number of targets being, for the former, 164, and for the latter 266.

TABLE II.

Class of shooters.	Distance in yards.	Total No. of arrows delivered.	Numbers of hits in the several colours.					Misses.
			Gold.	Red.	Blue.	Black.	White.	
Ladies	60	81696	1722	4927	7279	8572	8688	50508
Do..........	50	40848	1364	3799	5145	5640	5159	19741
Gentlemen	100	175968	1873	5365	8239	10629	11605	138257
Do..........	80	117312	2516	7061	10137	12067	12058	73473
Do..........	60	58560	2553	6651	8455	8983	7752	24166
Sums total		474384	10028	27803	39255	45891	45262	306145

(8.) To compare these results with theory, and ascertain how far the distribution of the hits in each series corresponds with our formula, the best way will be to deduce, in each, five separate values of a the constant appropriate to each, from—1st, the hits in the gold; 2d, the sum of those in the gold and red; 3d, the sum of those in the gold, red, and blue, and so on. These, it is manifest, in each series ought to agree *inter se*, though different for different series. Applying, then, the expression given in § (3.) for a to these entries of "hits" in Table II. on this principle, we derive corresponding values of our constant or modulus as in the annexed table (Table III.), in the first division of which it is set down in units and decimals, as in Table I.; while in the second are entered the same values reduced to inches and decimals by the multiplier 4·8.

TABLE III.

Class.	Distance in yards.	Values of a as deduced from the numbers of hits falling within the circles externally limiting the					Mean value of a.
		Gold.	Red.	Blue.	Black.	White.	
Ladies ……	60	5·704	5·715	5·777	5·867	6·003	5·813
Do………	50	4·518	4·530	4·631	4·751	4·882	4·662
Gentlemen	100	8·048	8·125	8·232	8·311	8·476	8·238
Do………	80	5·654	5·693	5·823	5·925	6·086	5·836
Do………	60	3·943	4·027	4·170	4·275	4·425	4·168
		in.	in.	in.	in.	in.	in.
Ladies ……	60	27·380	27·432	27·730	28·163	28·812	27·903
Do ………	50	21·685	21·743	22·231	22·804	23·432	22·379
Gentlemen	100	38·631	38·998	39·512	39·893	40·686	39·544
Do………	80	27·141	27·388	27·950	28·438	29,210	28·025
Do………	60	18·928	19·324	20·012	20·519	21·239	20·004

(9.) Although the general agreement of the results in each horizontal line of this table is quite sufficient to afford a highly satisfactory verification of the theory, it is impossible not to be struck with the uniform and steady increase (though small) in the value of a in proceeding from the gold outwards. There occurs not throughout the whole table a single instance in which this progressive dilatation is not maintained. For this there must be a reason; and the only rational account of it, so far as I can perceive, is this: Were the number of shooters infinite, including (indifferently) every gradation of skill, from absolute random shooting up to absolute certainty of striking the point aimed at: the result of their combination would be one from which individual skill would be entirely eliminated; and the distribution of the hits would be regulated purely and simply by the *intention* of hitting the centre. The *skill* in that case (on which the value of a depends) would be the average skill of the whole human race; and the value of a would conform to that average. But the number of shooters at each target being limited, never more than six, and sometimes not more than two or three (owing to our rejection of those scores where the arrows were not all delivered), the distribution of the hits on each results from the summation or superposition of several series of numbers graduated according to the values of so many different specified values of a as a modulus; the intermediate values being absent: and it by no means follows that such a series of sums will follow the law of gradation of either separately, or accommodate itself rigorously to an average modulus; any more than it would follow that a curve whose ordinate is the sum of the ordinates of (say), two concentric circles should itself be a circle. Suppose, for instance, we were to ground our calculation of an average *modulus* or value of a on a series of hits given by summing up each separate column of Table I. This would give the following series :—Gold, 862; Red, 1337; Blue, 1121; Black, 1042; White, 910, on a total number of 10,000 shots. And from this deducing five several values of a by the use of the formula of § (3), after the manner of those in Table III., we should obtain the series,

$$a = 2{\cdot}771,\ 3{\cdot}340,\ 3{\cdot}931,\ 4{\cdot}399,\ 4{\cdot}809,$$

which exhibits (only in a much more exaggerated form) precisely the sort of dilatation in question. This being then the tendency, in the case of each particular target, a similar tendency will of course be exhibited by the successive values resulting from a combination of any number.

(10.) The same principles apply of course equally to rifle-shooting as to archery, provided the target aimed at be circular. If rectangular, and especially if an elongated rectangle, the same formulæ will not apply; and the appropriate formulæ would be necessarily much more complex and their results proportionably more difficult of calculation. This is a strong argument for the use of circular targets: for, though for the mere decision of the order of merit in a distribution of prizes almost any impartial rule, rough and readily applicable, may suffice, the same cannot be said when the object is to obtain a true numerical measure of the national skill in the use of that great weapon: for which purpose it is highly desirable that the data afforded by our rifle prize meetings should be preserved, collected, and reduced systematically.

NOTE.

Demonstration of the formula in § (2.) *and* (3.)

The probability of committing the specific error r (*all errors presenting equal facility for their commission*) is proportional to $E\,(-kr^2)$, the characteristic sign E being used to denote the exponential or anti-logarithmic function; and k being some certain constant to be determined or eliminated. And in the case of aiming at the central point of a circular target, the degree of facility afforded for the commission of a lineal error r, no matter in what direction, is proportional to $2\pi r$, the circumference of a circle of that radius, or, simply to r: so that the probability of planting a shot somewhere *on the circumference* of that circle is measured by $r.\ E\,(-kr^2)$, and therefore the probability of making a hit anywhere *within its area* is proportional to $\int r dr.\ E\,(-kr^2)$ taken between the limits o and r. Representing certainty therefore by 1; this probability (which we have denoted by H in the foregoing pages) will be expressed numerically by such a multiple of this integral, so taken, as to give 1 for its value when r is infinite : which gives

$$H = 1 - E(-kr^2)\text{ ; and therefore } M = E(-kr^2).$$

Now, when $r = a$, the radius of that circle within which just half the total number of arrows strike ; the probabilities of hitting and missing that circle are equal : so that H and M are in that case each $= \frac{1}{2}$. We have, consequently,

$$E(-ka^2) = \tfrac{1}{2},$$

and eliminating k, we obtain

$$\text{Log. } M = \frac{r^2}{a^2}, \text{ log. } (\tfrac{1}{2})\text{ ;}$$

or,

$$M = \left(\frac{1}{2}\right)^{\frac{r^2}{a^2}}$$

COLLINGWOOD, *June* 8, 1866.

THE END.

www.ingramcontent.com/pod-product-compliance
Lightning Source LLC
LaVergne TN
LVHW021312110826
845150LV00003B/546

* 9 7 8 1 4 2 5 5 5 8 2 7 7 *